STATIONARY
ENGINEERING
HANDBOOK

STATIONARY ENGINEERING HANDBOOK

by K.L. Petrocelly, S.M.A./C.P.E.

Library of Congress Cataloging-in-Publication Data

Petrocelly, K. L. (Kenneth Lee), 1946-
Stationary engineering handbook / by K. L. Petrocelly.
   p.   cm.
   Includes index.
   1. Steam engineering--Handbooks, manuals, etc.
I. Title.
TJ277.P48   1989   621.1'6--dc19   88-45795
                                       CIP
ISBN 0-88173-078-5

Published by The Fairmont Press, Inc.
700 Indian Trail
Lilburn, GA 30247

Printed in the United States of America

10 9 8 7 6 5 4 3 2

ISBN 0-88173-078-5   FP

ISBN 0-13-844838-8   PH

While every effort is made to provide dependable information, the publisher, authors, and
editors cannot be held responsible for any errors or omissions.

Distributed by Prentice-Hall, Inc.
A division of Simon & Schuster
Englewood Cliffs, NJ 07632

Prentice-Hall International (UK) Limited, London
Prentice-Hall of Australia Pty. Limited, Sydney
Prentice-Hall Canada Inc., Toronto
Prentice-Hall Hispanoamericana, S.A., Mexico
Prentice-Hall of India Private Limited, New Delhi
Prentice-Hall of Japan, Inc., Tokyo
Simon & Schuster Asia Pte. Ltd., Singapore
Editora Prentice-Hall do Brasil, Ltda., Rio de Janeiro

Dedication

To my family who so graciously overlooked
all my bitching and moaning, . . . again.

Preface

Years ago, the only qualifications you needed to become an operating engineer were the ability to shovel large chunks of coal through small furnace doors and the fortitude to sweat profusely for hours without fainting. Since then, I'm happy to report, the profession has undergone a miraculous transition. As a consequence of technological evolution, the "engineer's" coal shovels have been replaced with computers and now perspiration is more the result of job stress than exposure to high temperatures. The domain of the operator has been extended far beyond the smoke-filled caverns that once encased him, out into the physical plant, and his responsibilities have been expanded accordingly. Unlike his less sophisticated predecessor, today's technician must be well versed in all aspects of the operation. The field of power plant operations has become a full-fledged profession and its principals are called Stationary Engineers.

Kenneth Lee Petrocelly

Contents

Chapter 1

THE PROFESSION

The field of Stationary Engineering and the position of Stationary Engineer are inextricably linked. Trying to separate them would be like taking sticky out of taffy. However, if you are to understand what each of the terms means, I must somehow find a way. I'll cover sticky here and taffy in Chapter 2. I hope you're pulling for me. I know, I know . . . sorry about that. Where is it written that dry subjects have to be dull?

WHAT IT IS

In a nutshell, Stationary Engineering is a profession involving licensed and unlicensed operating engineers, mechanics and laborers in the installation, operation, monitoring, inspection, maintenance, repair and disposition of physical plant equipment and systems. If saying that took your breath away, just wait until you hand-fire your first HRT on a hot summer afternoon . . . but more about that later.

Though not a science in and of itself, the field of Stationary Engineering draws upon and is structured around many sciences: mathematical, physical and biological. Once you leave the streets for a building's machinery spaces, you enter a wholly different world, the confines of which are only a mystery to most people; a place governed less by man's laws than those of nature; where matter, energy and their intricate intertwining are the only relevant factors. It's a training ground for fledgling plant managers, a repository for marine and locomotive engineers and a brotherhood for men of a particular breed.

Back in the days of yore or as my daughter so likes to put it, back in your days—which means a hundred years or so ago—there was no Stationary Engineering field. There was also little concern for the safety of equipment operators or the tenants in the buildings they served. There was no budget to adhere to, nor any idea how long a piece of equipment might last or that anything could be done to prolong its life. Overt fuel consumption wasn't a problem; the term energy conservation hadn't even been coined then. And equipment operators were often as unsophisticated as the machinery and vessels in their charge. It's beyond me how those could be referred to as the good old days.

Since then, the profession has evolved into a vanguard for industry, championing the preservation of equipment as well as human life and conservation of our limited natural resources. The professionals within its ranks have developed or adopted effective alternatives to energy waste, systems of education for its members and promotion of engineering economics in the physical plant. Today its main impetus is the cost-effective operation of electro-mechanical systems, the maximizing of equipment life and the safety of all persons involved in or affected by power plant operations.

WHERE IT'S PRACTICED

It would be simpler to list those places where the profession of Stationary Engineering isn't practiced than where it is; namely in log cabins nestled deep in the woods, on mass transportation vehicles such as cruise ships or airliners, and at nuclear testing ranges in the desert.

The fact is, almost no business or organization in the country can remain operational for long without the services of a person or group of persons trained in varying aspects of the field. Every building that is heated, cooled or ventilated, supplied with a utility service or contains equipment used for creature comfort or a manufacturing process, houses the profession. It is practiced in nuclear and fossil-fueled power generating plants, hospitals and nursing homes; in chemical company equipment yards, wastewater treatment plants, airport equipment rooms and cargo hangers; and in factories, schools and shopping malls. Anywhere you find a

Figure 1-1
Courtesy of Johnson Controls

production line, processing plant or place of public assembly, you'll find a worker plying the trade.

ELEMENTS OF THE JOB

The size of the plant and the extent of your duties within it determines the number of hats you will wear there. In a very large operation, each position may be its own specialty and effectively the hats are distributed one per man. Operators in medium-sized facilities are often called upon to wear more than one hat. Very small plants are sometimes manned by a single operating engineer who is responsible for the entire operation, especially after normal business hours. Haberdashers don't carry hat selections as large as his. Depending on which hat you happen to be wearing, at any given time in the power plant, you may be responsible for:

- heating, cooling and ventilating buildings
- generating electricity for local consumption
- starting, stopping and operating all types of electro-mechanical equipment
- sampling, analyzing and treating boiler and/or cooling waters
- using hand and power tools for making repairs to machinery, pressure vessels, piping, and valves
- preparing fuels for use in furnaces, internal combustion engines, heaters and processing units
- tracking utility consumption and energy conservation efforts
- troubleshooting equipment and system problems
- monitoring gauges, meters and other measuring instruments to assure proper operation of equipment
- adjusting throttles, valves, levers and switches to control flow
- observing and recording fluid levels, temperatures and pressures
- documenting equipment repair histories

- replacing worn bearings, belts, pulleys, gaskets and packing
- testing safety devices to ensure their operating integrity
- maintaining system components per manufacturer's instructions
- removing, calibrating and reassembling measuring instrumentation
- cleaning residues from equipment working surfaces
- lubricating rotating machinery and sliding devices
- measuring electrical circuits
- . . . etc.

Figure 1-2
Courtesy of Johnson Controls

Of course the forementioned hats we spoke of were referred to facetiously, but how you dress in machinery spaces must be carefully considered. Hazards and working conditions there will have a tremendous impact on what clothing and safety accoutrements you choose to attire yourself. Here are some things to look out for:

- heat from combustion and equipment surfaces
- escaping steam and hot water from leaks
- exhaust fumes resulting from the combustion process high noise levels
- slippery surfaces due to spilled oil, water or grease
- toxic fumes from solvents and other chemicals
- circulating smoke, dirt and dust due to inadequate air filtration
- flying particles from grinding wheels and hammer blows
- electrical shocks from exposed wiring and open control panels
- injuries from rotating or moving machinery parts
- explosions caused by over pressure conditions
- explosions caused by accumulations of unspent fuel
- skin burns from exposure to chemicals
- falls from ladders and equipment gratings
- burns from contact with inadequately insulated piping
- fires resulting from improperly stored combustibles
- cuts from protruding objects
- trips resulting from poor housekeeping
- ... etc.

IDENTIFYING THE WORK FORCE

The Stationary Engineer doesn't hold the only position in the profession but he holds the one job which is most representative of it and as such deserves a place of honor in this book. If you want to learn more about his/her responsibilities, you can skip to the next chapter, which is devoted entirely to this professional but it would be in your best interest as an apprentice to first view what's presented here.

Just as it is highly unlikely that a person can graduate from a university and immediately become CEO of an established organization, it's foolish to think that you can become a Stationary Engineer and take over the operation of a large power plant without first paying your dues. You may not hold every position available within its ranks on your way up, but the profession is a training ground and has its own unwritten apprenticeship requirements leading to journeyman status as a Stationary Engineer.

Depending on the type, size, needs, age and staffing of the power plant where you get your on-the-job experience, under what jurisdictional regulations it is governed and the policies of the company which owns it, you may find any or all of the following positions manned there:

- Boiler Operator
- HVAC Mechanic
- Engineman
- Maintenance Mechanic

Though these jobs are commonplace in the trade, by no means do they represent every conceivable position within, or operating aspect of, stationary engineering.

Over the years, external influences such as technological advances, engineering economics and the need to conserve energy, have caused many of the traditional jobs within the profession to become defunct or obscured. Position titles and job responsibilities vary not only from one jurisdiction to another but even between different plants in the same locale. What one company calls a technician, another may call a maintenance specialist, and persons having the same or similar titles may be charged with completely different job responsibilities. The intent here is not to provide you with a comprehensive classification of jobs found in the field, but rather a sampling of positions which best illustrates the work performed within it.

Here is a summary description of each of them:

Boiler Operator

Job Summary — Operates and maintains boilers, pressure vessels and auxiliary devices, equipment and their controls to produce steam and hot water for supplying power and heat for the comfort needs of building occupants and or use in manufacturing processes.

Figure 1-3
Courtesy of Johnson Controls

Working Conditions — Requires moderate physical strength and agility to bend, lift, climb, stoop, reach, squat, push and pull; prolonged periods of standing and visual observation of equipment; exposure to the hazards associated with steam and heat.

Required Skills — High school graduate or equivalent education, rudimentary mathematical ability, a basic knowledge of chemistry and physics, knowledge of boiler operating, cleaning and maintenance techniques.

Primary Duties — Start, stop and alternate equipment and systems to schedules or as the operation requires. Perform preventive maintenance on boilers and auxiliary equipment per established frequencies. Clean, and prepare boilers and pressure vessels for inspection. Repair and replace broken and/or worn machinery parts. Maintain operating logs on all equipment. Draw and test boiler water samples, recording the analysis and treatment given. Perform flue gas analysis and adjust air-fuel ratios at the burner.

Figure 1-4
Courtesy of Johnson Controls

HVAC Mechanic

Job Summary — Installs, operates, maintains and repairs refrigeration and air-conditioning equipment, systems, components, devices and controls to provide for the comfort needs of building occupants and/or conditioned air for use in a manufacturing process or storage of goods and foodstuffs.

Working Conditions — Requires repeated bending and stooping, climging of ladders and prolonged work overhead with arms in an extended position. Involves exposure to the hazards associated with electricity, mechanical action, gasses and toxic chemicals.

Required Skills — A diploma from a vocational or trade school certifying competency in the heating ventilation and air-conditioning field or equivalent on-the-job experience. A background in mathematics, physics, chemistry, basic electricity and blue print reading. Familiarity with the diagnostic tools of the trade.

Primary Duties — Start, stop and alternate refrigeration, heating and air-handling systems and equipment to schedules or as required by the needs of the plant. Perform prevention maintenance on compression and absorption refrigeration systems, boilers, hot water heaters and air-handling system components and control devices per established frequencies. Inspect, clean and repair and/or replace broken or worn machinery parts. Maintain operating logs on all equipment. Draw and test cooling water samples, recording the analysis and treatment given.

Engineman

Job Summary — Operates and maintains internal combustion engines, and electric motors used for driving electrical generators, refrigeration and air compressors, pumps and fans in the power plant.

Working Conditions — Required prolonged periods of time spent in awkward positions in close proximity to hot metal surfaces. Involves exposure to the hazards associated with combustible fuels, flammable liquids, electricity and toxic chemicals and fumes.

Figure 1-5
Courtesy of Johnson Controls

Required Skills — Vocational or trade school diplomas certifying competency in diesel and gasoline engine and electric motor repair. Basic background in mathematics, electricity and electro-mechanical devices. Familiarity with the diagnostic tools used in the trade.

Primary Duties — Start, stop and alternate engines and motors used to drive electro-mechanical devices to schedules or as required by the needs of the plant. Perform preventive maintenance on all equipment in his/her charge per established frequencies.

Inspect, clean and repair/replace worn and/or broken motor parts and mechanical power transmission devices such as couplers and gear boxes. Keep constant vigil on combustible fuel and flammable liquid qualities and operating levels. Maintain operating logs on all equipment and an adequate inventory of spare parts.

Figure 1-6
Courtesy of Johnson Controls

Maintenance Mechanic (Millwright)

Job Summary — Installs and repairs auxiliary equipment, physical structures and electro-mechanical devices located throughout the plant; not specifically maintained by others.

Working Conditions — Exposed to a wide variety of physical requirements and hazards. May be exposed to extremes of hot and cold and be called upon to work in inclement weather. Can work at very high heights or in the lowest levels of a building or structure. Is frequently exposed to a wide variety of physical hazards. May be called upon to lift heavy weights.

Required Skills — Should possess an associate degree in maintenance technology from an accredited trade school or equivalent licensure and possess proven skills in a number of disciplines, such as plumbing, welding, electricity, motor repair, fireman, pipefitting, or refrigeration. Good mathematical ability, a high mechanical aptitude, and a background in physics are a must.

Primary Duties — This position is a catch-all for any work not otherwise assigned in the plant. The maintenance mechanic works from a set of specific routines and is assigned repair work as problems manifest themselves throughout the physical plant.

No one of the prior listed positions is necessarily more important than another in its significance to the operation on the whole, as each is a specialty slanted in its own way to the overall operation and has its own basis for existence. You should receive training in aspects of all of them prior to having the designation of Stationary Engineer conferred upon you.

A PERSONAL OBSERVATION

Depending on whose scribblings you read, the profession had either a glorious or dubious inception. The truth be known, like any other aspiring field it had at its roots a bit of both, but mostly it just plodded along accepting its advances matter-of-factly as they came about. Stationary Engineering didn't abruptly burst onto the scene as a promising new field of endeavor; it evolved slowly, progressing as man progressed, one discovery or invention at a time, unbiased and unprejudiced.

Much of the history written concerning it didn't deal with the field itself, but with the societies and other organizations which attached themselves to it. Many of those entering the record in that era were less historian than opinionated and since I wasn't there, I put little stock in what they had to say; nor should you. However, there is no denying that the organizations did and do exist and that they have had and do have a profound effect on the profession—always for its betterment, never to its detriment—as the impetus of the profession is progressive and blind to failure or circumstance. The profession found its beginnings in the thoughts of engineers and the sweat of laborers from a by-gone age and was organized by the brotherhoods they spawned. We owe much to those who went before us.

Personally, I'm concerned less with yesterday's practices than I am with tomorrow's direction. Who, in hell's name, wants to go through that again? The future of Stationary Engineering holds the promises of personal achievement, higher wages and professional recognition for those of us who are able to match the pace set by the profession.

Chapter 2

THE OPERATING ENGINEER

I'm not suggesting that the Stationary Engineers are the only professionals to be found in our ranks; many positions require licensure, but the title alone is a proclamation that its bearer reigns unchallenged over the limited kingdom he/she surveys. Of course, that is why wanting to become a Stationary Engineer is such a noble aspiration. But before you can ascend to that lofty height, there are some things you need to know . . .

WHO HE IS

We probably shouldn't progress any further without addressing the he/she/person thing that inevitably confronts us when writing on subject matter such as this. Women have always been welcome as visitors to our sanctuaries and recently we have even accepted her as an equal and made room for her on the throne. But on the basis of majority rule, it is man who presently dominates this field; so for purposes of this work, all persons active in the profession will herewith be referred to in the male gender.

As just ascribed, it's probable the next Stationary Engineer you meet will be a man. Chances are, at least at this writing, he will have received his initial training as a locomotive engineer, merchant marine or while on duty with the U.S. Navy in the engineering section. It's highly possible his age will range between 40 and 65 and that he served in the military during the Second World War, in Korea or Vietnam. If that is the case, in all likelihood he'll also be wearing a green or khaki uniform.

The foregoing describes the traditional operating engineer; the old guard if you will. Most of his expertise was received on the job from those preceeding him, his own experiences and what he read. In contrast, the new breed of Stationary Engineer is book trained in the classroom and moves more quickly through the various skill levels of the profession. Apprenticeship is a foreign word to him but he would do well to learn a few things from us dinosaurs before our collective extinction, though to their misfortune, many won't even make an attempt.

Whether you pledge your allegiance to and live by the traditions of the brotherhoods that got us here or have chosen to ally yourself with the new regime, is of no consequence once work begins. As long as we are divided into old and new factions, disagreements between us will abound. It might be difficult to admit, but under the skin we really are quite alike. So, if our individual professional committments are on a par, where do our differences lie? My guess is, it doesn't matter. When it gets right down to it, we are all of the same cloth. What's important is, whereas our philosophies and methods may differ drastically, our primary objectives remain identical: to provide professional care for the equipment in our charge and to ensure the comfort and safety of all persons affected by our operating decisions.

HOW HE'S GROWN

In the not-too-distant past, a Stationary Engineer's job was, albeit physically demanding, rather simple. He tended coal-fired furnaces, maintained proper water levels and steam pressures in boilers and made certain that his reciprocating pumps and engines were kept lubricated. Today, many of us fire dual- and tri-fuel boilers, are responsible for monitoring sophisticated operating controls and adhere to rigid preventive maintenance schedules in caring for our equipment. No longer are we relegated to the bowels of our buildings strictly to shovel coal and haul ashes. Our capabilities are matched to today's technology. We take on ever-increasing responsibilities within our profession and educate ourselves in trades traditionally foreign to us. Instead of being just producers of steam and fixers of machines, we take a holistic approach to operating our power plants. Today we are often made responsible for every aspect of utility service and systems operation in our buildings.

WHAT HE NEEDS TO KNOW

A Stationary Engineer's competence is reflected by his proven abilities on the job and his knowledge of power plant operations as certified by his license. Due to the complex and diverse nature of his work, he must at once be both genius and genie; for when problems arise in the power plant he is expected to know their solutions and when logical answers aren't readily available, he must use his common sense to surmise them. How much of each he must possess is affected by the age, size and condition of his plant, the ordinances of the municipality in which it resides, and the standard operating prodecures of the company who owns it.

Standardization of rules regulating the requirements for educating, examining and licensing Stationary Engineers is a constant source of frustration for employers trying to lower their operating costs and governing bodies charged with maintaining the safety of the public. Over the years, attempts have been made by many different organizations and state legislatures to compromise on the issues, but inevitably their efforts failed. One of the more successful groups in this regard is the National Institute for the Uniform Licensing of Power Engineers, Incorporated. The N.I.U.L.P.E. is a third-party licensing agency that acts on a national level to establish standards for firemen and water-tenders, engineers, operators, examiners, instructors and the licensing agencies currently existing. In addition, they will accredit courses taught in Power Engineering Technology which meet minimum requirements and will commission those instructors of power technology and those instructors teaching courses in support of power technology who meet the requirements established by this Institute. At last count more than 30 states subscribe to their tenets in some form. A more complete description of this organization is given in Appendix A in the rear of this book.

Just as jobs vary in magnitude, one to the other, so do the job skills needed to accomplish them. Whereas a Stationary Engineer's license is indicative that its holder is familiar with the practice of Stationary Engineering, it is the classification of that license which determines to what degree. Or to put it in words that an employer would understand . . . why pay a first-class operator at twice the hourly rate when a fourth-class engineer is qualified to handle the work? Conversely, a third-class operator shouldn't be substituted when the expertise of a chief engineer is needed on the job. The bottom line, as always, is that you get what you pay for. The classification system provides employers with personnel whose degree of skill can be matched to the job and provides employees with a vehicle for promotion through the ranks within the profession.

The N.I.U.L.P.E. classification system includes the following qualifications and curriculum requirements in ascending order of skill level.

FOURTH-CLASS ENGINEER

Qualifications

Minimum age – 18

Education Required – Two years high school, 2 years approved apprenticeship, or approved OJT

Experience Required – Two years minimum

Examination Required – Oral, Written, Practical

Maximum Engine (Prime Mover) Horsepower requiring licensed power engineer (Unsupervised) – 500 HP

Maximum Boiler Horsepower requiring licensed power engineer (Unsupervised) – LP-150 HP; HP-25 BHP

Refrigeration License (Unsupervised) – 100 tons

Suggested Curriculum

- **Water glass**: Placement, function, how held in place, length, maintenance, valves needed, replacement, how and when tested.

- **Water column**: Placement, function, how held in place, length, care, what could affect its efficiency, maintenance, valves and cocks needed, how and when tested.

- **Safety valve and rupture discs**: Placement, function, care, maintenance, how and when tested, description, potential failures, adjustment for pressure.

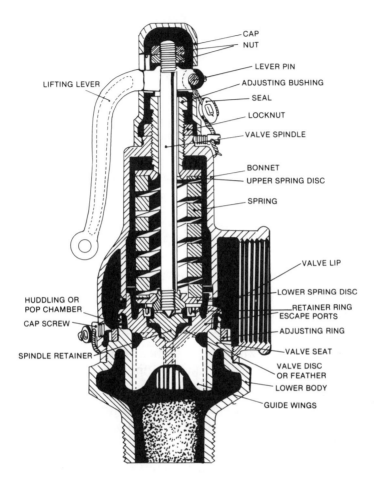

Figure 2-1. Safety Valve
(Courtesy: Building Owners and Managers Institute)

- **Steam gauge and siphon**: Placement, principle, function, care, maintenance, potential failures, how and when tested.
- **Feed water, piping and valves**: Location, care, maintenance, potential failures, reason for valves on piping, how kept in good condition, reason for internal pipe.
- **Blow down valves, piping and tank**: Location, purpose, potential failures, care.

- **Fusible plug**: Where placed, purpose, how kept in good condition, possible failures, description, installation, when renewed.

- **Stays**: Placement, purpose, types, description, care, advantages of various types, potential failures.

- **Two common types of boilers — fire tube and water tube**: Description and characteristics, qualities of a good boiler.

- **Dangerous conditions**: Whan a boiler should not be operated, causes of boilers being burned or exploded.

- **Corrosion, pitting, priming and foaming, bulging, bagging**: Meaning of each, where found and how created, dangers and remedies.

- **Scale and mud**: Where found, cause, prevention, effect, dangers, removal.

- **Feed water treatment**: Meaning, purpose, how applied and controlled, dangers of over-treatment and under-treatment.

- **Feed water heaters**: Types, purpose, advantages, methods of heating feed water, applications, potential failures.

- **Mathematics**: A knowledge sufficient to solve any simple problem involving division and multiplication.

- **Elementary combustion**: Mixtures of combustibles and air, methods of application and how controlled, purpose of setting, dampers, draft, chimneys or stacks, hand firing methods, heating values of anthracite or bituminous coal, oil and gas.

- **Pumps — simplex, duplex, vacuum**: Care, maintenance, purpose, potential failures, remedies, description.

- **Injector**: Function, principle, care, maintenance, cleaning and inspection, how often, purpose, valves.

- **Cleaning and inspection of boilers and setting.**

- **Starting clean boiler**: Manhole covers, how replaced and removed, raising steam, cutting into live steam header, banking boiler and starting after bank.

- **Grate surface, fire line, water level, fire tubes, water tubes, heating surface**: Meaning of these terms.

- **Operation of stoker**: Starting, purpose, maintenance, care, potential failures, advantages and disadvantages.

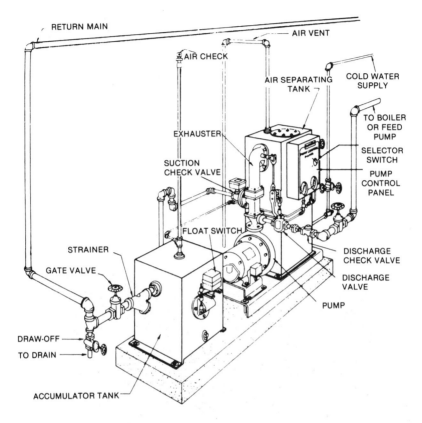

Figure 2-2. Vacuum Pump
(Courtesy: Building Owners and Managers Institute)

- **Operation of oil or gas burners and electric boilers:** Installation, starting, care, possible failures, controls, safety devices.

- **Steam non-return valve, expansion joints, heaters, steam separator, sight feed lubricator, steam trap reducing valve, feed water regulator:** Function, use and location.

- **Refrigeration compressor, condenser, receiver, evaporator, purge, expansion valve, charting, liquid, suction, discharge, cross-over valves:** Purpose, location, dangers, correct operating procedures.

- **Air compressor:** Dangers in operation, maintenance, correct operating procedure.

- **Electrical equipment**: Fuses, cut outs, relays, switches, circuit breakers, purposes and comparative applications, dangerous conditions in operation of a motor, prevention of starting, sizes of fuses, carrying capacities of wires for lighting circuits, volt, ampere, watt, Ohm, D.C., A.C., electrical conductor, electrical insulator: meaning of these terms, difference between electric generator and motor.

- **Steam Engines**: Types, setting valves, purpose of fly wheel, eccentric, governor, cross head, lead, lap, angle of advance, valve travel and cut off, methods of lubrication and application to various parts, how started and how shut down, maintenance and care.

- **Steam Turbines**: General knowledge of the lubricating system, governors, and throttle valves.

Figure 2-3. Deposits on turbine blades
(Courtesy: Betz Laboratories)

- **Steam condensers**: General knowledge of condensers, where, how and why they are used, general care and upkeep of condenser auxiliaries.

- **Steam plant accessories**: Back pressure valves, non-return valves, throttle valves, expansion joints, feed water regulators, steam separators, sight feed lubricators, steam traps, reducing valves, sprinkler systems, function, location, operation and care.

- **Heating, air conditioning and ventilation**: Methods, controls, meaning of water hammer, piping arrangements, radiation, vacuum and plenum systems in mechanical ventilation, gravity and vacuum steam systems, maintenance.

- **Operation and maintenance of controls.**

- **Air pollution and ecology.**

- **Plant safety.**

THIRD-CLASS ENGINEER

Qualifications

Education Required — High school graduate or GED, 3 years approved apprenticeship, or approved OJT
Experience Required — 3 years minimum
Examination Required — Oral, Written, Practical
Minimum time in previous grade — 1 year
Maximum Engine (Prime Mover) Horsepower requiring licensed power engineer (Unsupervised) — 1000 HP
Maximum Boiler Horsepower requiring licensed power engineer (Unsupervised) — LP-Unlimited; HP-200BHP
Refrigeration License (Unsupervised) — 500 tons

Suggested Curriculum

- **Water glass**: Placement, function, how held in place, length, maintenance, valves needed, replacement, how and when tested.

- **Water column**: Location, function, how held in place, length, what could affect its efficiency, maintenance, valves and cocks needed, how and when tested.

- **Safety valve and rupture discs**: Placement, function, care, how and when tested, description, what could go wrong, adjustment for pressure.

- **Steam gauge and siphon**: Where and how placed, principle, function, maintenance, failures, how and when tested.

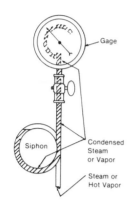

Figure 2-4. Syphon Operation
(Courtesy: Building Owners and Managers Institute)

- **Feed water, piping and valves:** Location, care, how kept in good condition, potential failures, reason for valves on piping, valve maintenance, reason for internal pipe.

- **Blow down valves, piping and tank:** Placement, purpose, maintenance, failures.

- **Fusible plug:** Location, purpose, maintenance, potential failures, description, installation, when renewed.

- **Stays:** Where placed, purpose, types, description, care, advantages of various types, potential failures.

- **Two common types of boilers — fire tube and water tube:** Description and characteristics, qualities for a good type of boiler.

- **Dangerous conditions:** When a boiler should not be operated, causes of boilers being burned or exploded.

- **Corrosion, pitting, priming and foaming, bulging, bagging:** Meaning of each, where found and how created, dangers and remedies.

- **Scale and mud:** Where found, cause, prevention, effect, dangers, removal, reduction.

- **Feed water treatment:** Meaning, purpose, how applied and controlled, dangers of over-treatment and under-treatment.

- **Feed water heaters:** Types, purpose, advantages, methods of heating feed water, applications, failures.

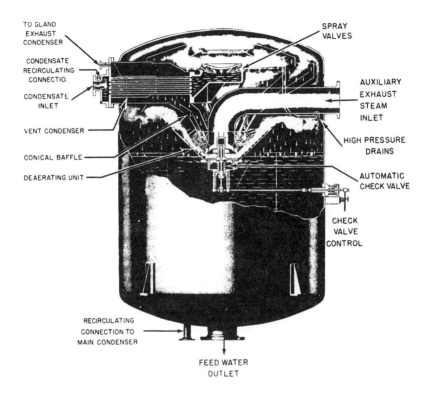

TO GLAND
EXHAUST
CONDENSER

CONDENSATE
RECIRCULATING
CONNECTIO.

CONDENSATE
INLET

VENT CONDENSER

CONICAL BAFFLE

DEAERATING UNIT

SPRAY
VALVES

AUXILIARY
EXHAUST
STEAM
INLET

HIGH PRESSURE
DRAINS

AUTOMATIC
CHECK VALVE

CHECK
VALVE
CONTROL

RECIRCULATING
CONNECTION TO
MAIN CONDENSER

FEED WATER
OUTLET

Figure 2-5. Deaerating Feed Tank

- **Mathematics**: A knowledge sufficient to solve any simple problems involving division, multiplication, decimals, measurement, ratio and proportion.

- **Elementary combustion**: Mixtures of combustibles and air, methods of application and how controlled, purpose of setting draft, chimneys or stacks, hand firing methods, heating value of anthracite and bituminous coal, oil and gas, description.

- **Pumps — simplex, duplex, vacuum rotary, centrifugal**: Maintenance, purpose, potential failures, remedies, description.

- **Injector**: Function, principle, care, maintenance, cleaning and inspection, how often, purpose, valves.

- **Cleaning and inspection of boilers and setting.**

- **Starting clean boiler**: Manhole covers, how replaced and removed, raising steam, cutting into live steam header, banking boiler and starting after bank.

- **Grate surface, fire line, water level, fire tubes, water tubes, heating surface**: Meaning of these terms.

- **Operation of stoker**: Starting, purpose, maintenance, what could go wrong, advantages and disadvantages.

- **Operation of oil or gas burners and electric boilers**: Installation, care, failures, controls, safety devices.

- **Steam non-return valve, expansion joints, heaters, steam separator, sight feed lubricator, steam trap reducing valve, feed water regulator**: Function, use and location.

- **Refrigeration compressor, condenser, receiver, evaporator, purge, expansion, charging, liquid, suction, discharge, crossover valves**: Purpose, location, dangers, correct operating procedures.

- **Heat engines**: General knowledge of types — steam, internal combustion, turbines.

- **Air conditioning**: General knowledge of fundamentals, temperature and humidity control, principles of operation.

- **Air compressor**: Dangers in operation, maintenance, correct operating procedure.

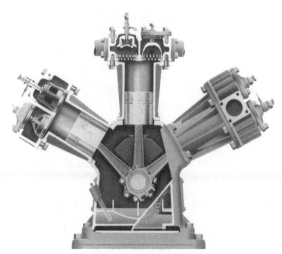

Figure 2-6. Air Compressor
(Courtesy of Ingersoll-Rand)

- **Electrical equipment**: Fuses, cut outs, relays, switches, circuit breakers, purposes and comparative applications, dangerous conditions in operation of a motor, prevention of starting, sizes of fuses, carrying capacities of wires for lighting circuits, volt, ampere, watt, Ohm, D.C., A.C., electrical conductor, electrical insulator – meaning of these terms, difference between electric generator and motor.

- **Operation and maintenance of controls.**

- **Air pollution and ecology.**

- **Plant safety.**

SECOND-CLASS ENGINEER

Qualifications

Education Required – High school, 4 years approved apprenticeship, or approved OJT

Experience Required – 4 years minimum

Examination Required – Oral, Written, Practical

Minimum time in previous grade – 1 year

Maximum Engine (Prime Mover) requiring licensed power engineer (Unsupervised) – 2500 BHP

Maximum Boiler Horsepower requiring licensed power engineer (Unsupervised) – 500 BHP

Refrigeration License (Unsupervised) – 1000 tons

Suggested Curriculum

- **Draft**: What it is, how created, purpose, types, controls, measurement, factors affecting draft, purpose of chimney, amount of draft created by chimney per 100′ of height.

- **Elementary combustion**: Difference in heating value between anthracite and bituminous coal, oil, gas and coke, cause and prevention of baffles, bridge walls, combustion chamber, cause of clinkers, how reduced and how removed, maintenance and repair.

- **Soot**: Effect, prevention and removal, operation of soot blowers, loss of efficiency due to soot.

- **Types of boilers**: Construction details, how installed in setting, purpose of each and where used to best advantage, super-heaters, air preheaters, economizers, dry pipes, internal baffles, fusible plugs, type and where installed, types of internal feed pipes, steam and safety valve outlets, caulking tools, type cutters, expanders and beading tools.

- **Stays**: Where used, purpose, types, description, care, advantages and disadvantages of types.

- **Boiler joints**: Types, riveting and welding procedures, where used, relative strength of each type and various locations, how caulked, what could go wrong, remedies, testing tightness of rivets, meaning of single and double sheer.

- **Boiler and furnace**: Washing of shell and tubes, cleaning and inspection, how often, purpose, valves, fittings and settings, description of methods and parts checked, tube, shell scrapers, and brushes, air and water turbining tools.

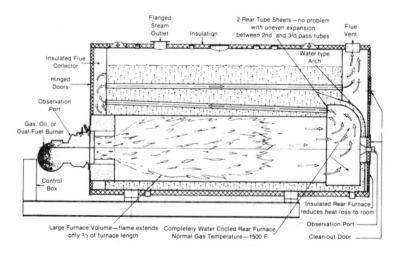

Figure 2-7. Three Pass Firetube Boiler
(Courtesy: Building Owners and Managers Institute)

- **Corrosion, pitting, grooving, fire cracks, bulging, bagging, blistering**: Causes, dangers, prevention and remedies.

- **Feed water piping and valves**: Location, care, typical installations.

- **Scale**: What it is, how it gets into boiler, dangers of too much, how prevented and how removed, where it is found.

- **Feed water treatment**: Methods, how applied and controlled, knowledge of impurities in feed water and their effect on a boiler under pressure and temperature, how to reduce these effects, priming and foaming, meaning, causes and remedies, purpose of continuous blow down and when used, purpose of main blow down and when used.

- **Feed water heaters**: Types, purposes, locations, advantages, disadvantages, maintenance, potential failures.

- **Water glass and column**: Maintenance, what could go wrong with either, importance as a safety factor in operating.

- **Steam gauge and siphon**: Principle of operating, where placed, care, failure, how and when tested, importance as a safety factor in operating.

- **Safety valve, relief valves, and rupture discs**: Purpose, care, how and when tested, description, potential failures, adjustment for blow down, reason for chatter, how drains and outlet piping installed.

- **Injector**: Principle, purpose, construction, care and maintenance, how connected, malfunctions.

- **Operation of stokers**: Types, how started, how controlled, advantages, maintenance, how ashes are removed.

- **Operation of an oil or gas burner and electric boilers**: Starting, stopping, controls, safety devices, types of oils in burning installations and how used, types of burners, dangers connected with oil-burning installations, care of automatic equipment and knowledge of what could go wrong.

- **Description of simplex, duplex, rotary, centrifugal and vacuum pumps**: Starting a feed pump, knowledge of proper operation, setting valves in steam pumps, lubrication, packing used and how to apply on all applications connected with steam feed pumps.

- **Steam engines**: Types, setting valves, purpose of fly wheel, eccentric, governor, cross head, lead, lap, angle of advance, valve travel and cut off, methods of lubrication and application to various parts, how started and how shut down, maintenance and care, dangers.

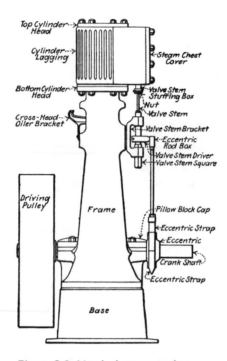

Figure 2-8. Vertical steam engine.

- **Steam turbines**: General knowledge of the lubricating system, governors and throttle valves.

- **Steam condensers**: General knowledge of condensers, where, how, and why they are used, general care and upkeep of condenser auxiliaries.

- **Mathematics**: Sufficient knowledge to enable candidate to work problems involving multiplication, division, decimals, measurement, square roots of numbers, powers of numbers, reciprocals, ratio and proportion, and formulas.

- **Steam plant accessories**: Back pressure valves, non-return valves, throttle valves, expansion joints, feed water regulators, steam separators, sight feed lubricators, steam traps, reducing valves, sprinkler systems—function, location, operation, care.

- **Heating, air conditioning and ventilation**: Methods, controls, meaning of water hammer, piping arrangements, radiation, vacuum and plenum systems in mechanical ventilation, gravity and vacuum steam systems, maintenance.

- **Refrigeration**: Starting, operating, cycle, purpose of safety head, charging, expansion, suction, discharge, cross over, purge, liquid oil valves, purpose of condenser, receiver, compressor, oil separator and evaporator, types of condensers, how ammonia is added and removed, how air is removed, how to defrost, effect of excess oil, how to add oil, type of oil, type of packing, how to pack stuffing boxes, purpose of brine pumps.

- **Air compressor systems**: How to start, how to operate, details of cooling system and how controlled, details of lubrication, danger in operating air compressors, controls, filters and safety devices.

- **Electrical equipment**: Definitions of switches, rheostats, circuit breakers, disconnects, rotors, stators, armatures, commutators, slip rings, two- and three-pole switches, volts, amperes, watts, ohms, lamps in series, lamps in parallel, how current flows, definitions of magnetism, induction, frequency, meaning of a short circuit and how it is caused, meaning of a ground wire, how electric current measured, explanation of how two wires are formed into a joint, calculations for computing sizes of fuses in lighting and motor circuits, dangers connected with the operation of a motor or generator.

- **Operation and maintenance of controls.**

- **Air pollution and ecology.**

- **Plant Safety.**

FIRST-CLASS ENGINEER

Qualifications

Education Required — High school, technical school, job experience

Experience Required — 6 years minimum

Examination Required — Oral, Written, Practical

Minimum time in previous grade — 2 years

Maximum Engine (Prime Mover) horsepower requiring licensed power engineer (Unsupervised) — 7500 HP

Maximum Boiler Horsepower requiring licensed power engineer (Unsupervised) — 1500 BHP

Refrigeration License (Unsupervised) — 5000 tons

Suggested Curriculum

- **Steam boilers and generators**: Types of boilers, principles, advantages and disadvantages, steam, electric cylindrical, vertical, locomotive, water tube, fire tube, packaged boilers, flash boilers and scotch marine boilers, steam accumulators, externally fired, internal-fired boilers, heating surfaces, how computed, types relative to horse power rating, super heaters, pre-heaters, economizers, dry pipes, internal baffles, where and how installed and purpose of each.

Figure 2-9. Package watertube boilers.
(Courtesy: Betz Laboratories)

- **Boiler construction**: Various forms, plates, rivets and welding used, meaning of shearing stress, tensile stress, modulus of elasticity, bursting pressure, working pressure, factor of safety, tensile strength, meaning of pitch percentage of plate strength, percentage of rivet strength, unit section of plate, how to calculate strength of boiler heads, shells, efficiency of joints, stays, description and sketches of joints, relative strength, percentage of rivet strength, unit section of plate, how to calculate strength of boiler heads, shells, efficiency of joints, stays, description and sketches of joints, relative strengths of various joints and how computed. How to calculate combined efficiency of rivetted joints, computation of working pressure-given, tensile, strength, factor of safety, thickness, and radius of shell; also efficiency of joint, calculations for the tube ligaments and dished heads, staying of flat surfaces, types of stays, limitation of stress and allowances for various types of stays, constants for different forms and applications of stays, doubling plates, how to calculate the area of the heads or segments of boilers.

- **Boiler settings**: Types, purpose, description, radiation, bridge walls, baffles, ignition, mixing and deflection arches, combustion chambers, furnaces, meaning of sufficient height over grates, effect of combustion space on capacity and efficiency, how to build or repair bridge walls and setting, sketches of typical settings, materials and cements used.

- **Inspection and testing**: Washing of shell and tubes, cleaning and inspection, how often, purpose, valves, fittings and setting, description of method and parts checked, what to look for, possible defects and how corrected, hydrostatic tests and how applied, purpose and value.

- **Types of steam power plants**: Simple, non-condensing, elementary, turbo-alternator plants, central stations, high pressure plants, super pressure plants (power), distribution and losses in each type.

- **Fuels**: Classification of wood, coals, gas, oil and waste heat, coal sampling and analysis, comparative heat values of various fuels, storage of coal, a knowledge of the methods used in handling coal and ashes, losses in handling, cause, prevention and removal of clinkers.

- **Combustion**: The combustion process and reactions, theory and practice, air requirement, supply and control, draft measurement and control, description and sketches of various draft gauges, dampers, purpose and controls, amount of draft needed under and over fires, balanced draft and its practical meaning, description of flue gas analyzer, how and where used, combustion efficiency and how obtained, temperatures in furnace and breeching, excess air and its effect on efficiency, advantages of forced and induced draft, air pollution, transfer of heat to boiler water, how heat is transferred.

- **Mechanical stokers**: Types, principles, purpose, description, comparison, how operated, how maintained and repaired, controls, what could go wrong, remedies.

- **Oil and gas burner installations**: Types of burners, type of oils used in burning installations and how used, dangers connected with burning installations, how to operate and maintain automatic equipment, make sketch of one type each of industrial burner for gas and oil, a sketch of piping, valves and allied equipment for an oil-fired packages boiler.

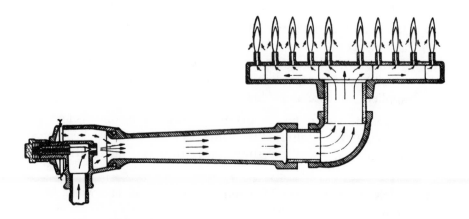

Figure 2-10. Atmospheric Burner
(Courtesy: Building Owners and Managers Institute)

- **General knowledge of a simplex, triplex, weir, rotary centrifugal and duplex or vacuum pump**: Knowledge of proper operation, where installed, lubrication, types of water pistons, performance, packing uses and how applied, what could go wrong, how to calculate size of pump for a feed line to boiler rating and over rating.

- **Piping**: Computation of sizes for water and steam, how installed, allowance for drainage and friction, flow of steam, Napiers rule, sizes of piping for pump discharge, also for feed line to boiler.

- **Feed water heaters**: Types, sketches, saving fuel, deaerators, trays, filters, oil separators, back pressure valves, position of various types, water measurements, temperatures and pressures carried, water seals, possible malfunctions.

- **Feed water conditioning**: General knowledge of the chemical changes that occur when feed water is heated in a boiler. Scale, what it is and what causes it, danger of too much, impurities generally found in raw water, carry over, embrittlement, corrosion, methods of conditioning, internal and external treatment. Hot and cold lime soda, Zeolite, demineralization, ion and anion exchangers, supplementary phosphate treatment, caustic soda and its application. Water analysis, tests, parts per million, grains per gallon, soap hardness test, alkalinity test (PM and PH), chloride, sulfite use of chelants, and phosphate tests, total dissolved solids.

- **Steam engines**: General knowledge of types—simple, compound, triple expansion, quadruple, uniflow, Corliss, high-speed slide valve, specific knowledge of one type of engine, governors, throttle, inertia, centrifugal, shaft, sketches of governors, types of valves, how to set valves on various types of engines, methods of lubrication and application to various parts, how started and how shut down, maintenance and care, dangers, indicators, indicator diagrams and how to read.

- **Steam turbines**: General knowledge of how turbines convert heat energy into mechanical energy and able to describe the bearing, oiling system, governors, and auxiliaries used with turbines, specific knowledge of dangers relative to starting and stopping a steam turbine.

- **Steam condensers**: General knowledge of the construction of the different types of condensers, together with a knowledge of their auxiliaries and advantages gained by the use of condensers on various types of engines and turbines.

- **Steam plant auxiliaries**: Superheaters, economizers, air pre-heaters, feed water regulators, feed water heaters, temperature regulating valves, flow meters, return thermostatic and bucket traps, back pressure valves, non-return valves, safety valves, reducing valves, boiler feed pump governors, air unloaders, expansion joints, steam separators, oil burners, calorimeters, ejectors, siphon, sight feed lubricators, descriptions and sketches of each.

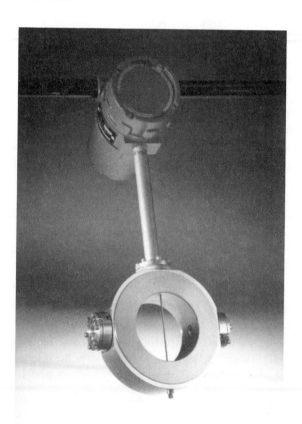

Figure 2-11. Flowmeter.
(Courtesy J-TEC Associates, Inc., Cedar Rapids, IA)

- **Definitions**: British thermal unit, foot pound, latent heat, specific gravity, specific heat, mechanical equivalent, inertia, convection, matter, momentum, siphon, clearance pockets, intercooler, aftercooler, horsepower of boiler, horsepower of engine, ton of refrigeration, axial flow, dynamic static, alkaline, freon, propane, atomic weight, barometer, hydrometer, Boyles Law, capillary, caustic soda, oscillation, center of gravity, centrifugal force, cohesion, velocity, speed, acceleration, conversion ($F°$ - $C°$ - $R°$), mean effective pressure, density, back pressure, oxidation, thermal efficiency.

- **Mathematics**: Knowledge of fractions, decimals, percentage and square root, knowledge of how to find the area and circumferance of a circle, also the volume of a cylinder, understanding of the relationship between the sides of right angle triangles, knowledge of the method of solving problems relating to a chain of gears, pulley sizes, rope blocks, screw jacks, levers, etc., understanding of the factor of evaporation, equivalent evaporation, latent heat, total heat, and a definite understanding of the problem involving the abstraction or supply of heat, relation to heating water or making ice, how to solve problems in heat supplied and heat extracted and how to compare boiler performances, how to solve problems involving heat calculations of steam production, boiler efficiency and horsepower, how to find the horsepower of an engine from an indicator diagram, how to solve problems involving specific gravity, calculations involving steam pressure on areas, how stays are calculated and designed, how to solve problems involving costs of operation relative to fuel and labor costs, how to calculate heating surface and horsepower of various types of boilers.

- **Heating, air conditioning and ventilation**: Methods, controls, piping installations, radiation, vacuum and plenum system in mechanical ventilation, gravity and vacuum steam systems, design, operation and maintenance.

- **Refrigeration**: Cycle theory of how heat is abstracted, common refrigerants and their properties, direct and indirect systems, purpose, location and sketches of safety head, condenser, receiver, compressor, oil separator and evaporator, types of condensers, how to operate refrigeration systems, how to charge, reduce or remove refrigerant from system, changing or adding oil, pumping out, condensing and defrosting, dangers in operating, annual inspection, why cooling water on compressor and condenser, how ice is manufactured, how ammonia leak is found and stopped, refrigeration capacity and losses, how air is removed, types of oil, how added, effect of too much, types of packing, how applied, purpose of brine pump, controls, absorption systems, hand and automatic, safety valves and discharge pipes.

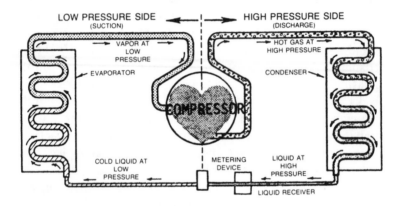

Figure 2-12. The Refrigeration Cycle
(Courtesy: Building Owners and Managers Institute)

- **Air compressor systems**: How to start, how to operate, details of lubrication, purpose of unloaders, description and sketches, compound compressors, intercoolers, aftercoolers, air receivers, controls, filters and safety devices, inspection, dangers in operating air compressors.

- **Electricity**: Meaning of terms—cycle, phase, starting torque voltage drop and voltage fluctuation, excitation, motor clearance, electrical horsepower, auto transformer, starter taps, disconnect, inductance, collector rings, slip, commutator, switch board, ground wires, power factor, peak load, magnetizing current, circuit breaker, no voltage release, overload relay, transformer, bus bars, dynamometer, frequency reactance, resistor, synchronism, brush rocker, EMF, impedence, interpole, lag, motor convector, rheostat, description of main types of A.C. motors, characteristics and applications, how these various types are started and controlled, connection or wiring diagrams for each type, description of auto transformer starter, also of magnetic contactor starter, and of a triplex rheostat, care and maintenance of electrical machinery and equipment, troubles, testing, and correction.

- **Lubricating oils:** Knowledge of the ingredients and qualities of oils, how they are filtered and applied, storage of oils.
- **Chemistry and physics:** Basic fundamentals as applied to power plant theory and practice.

CHIEF ENGINEER

Qualifications

Education Required – High school, technical school, plant and management experience
Experience Required – 10 years
Examination Required – Oral, Written, Practical
Minimum time in previous grade – 4 years
Maximum Engine (Prime Mover) Horsepower requiring licensed power engineer (Unsupervised) – Unlimited
Maximum Boiler Horsepower requiring licensed power engineer (Unsupervised) – UNLIMITED
Refrigeration License (Unsupervised) – UNLIMITED

Suggested Curriculum

- **Fuels:** Classification of wood, coal, gas, oil and waste heat, composition of coal, coal sampling and analysis, proximate and ultimate analysis, Dulongs formula, comparative heat values of various coals and other fuels, storage of coal and other fuels, losses in handling, advantages and disadvantages of each type of fuel.
- **Combustion:** Fundamentals of the chemistry of combustion, stages of combustion, the combustion process and reactions, combustible constituents of various fuels, the incombustible for complete combustion, theoretical and actual weight of

air required per pound for combustibles, flue gases and analysis, the flue gas analyzer apparatus, automatic flue gas recorder, combustion efficiency and how obtained, excess air and its effect on efficiency, value of CO_2, CO and O readings, air requirements, supply and control, temperatures, measurement and controls in furnace, passes and breeching, smoke, cause, measurement and prevention, smoke density recorder Ringlemann chart, losses due to flue gas, combustible in ash, moisture in fuel, moisture from combustion of hydrogen, moisture in air and radiation plus heat absorbed by boiler or a complete heat balance.

• **Draft**: Natural, chimneys, stacks, types and determination of size, empirical chimney formulas, mechanical draft, induced and forced draft, Prat, Evase or Thermix systems, balanced draft systems and controls, measurement of draft, descriptions and sketches of various draft gauges, amount of draft needed under and over fire (in passes and stack in various settings and types of fuel and the methods of burning the fuel), fans—types and controls, breechings and dampers systems—description and sketches.

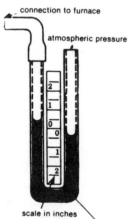

Figure 2-13. U-Tube Draft Gage
(Courtesy: Building Owners and Managers Institute)

• **Steam boilers and generators**: Types of boilers—waste heat, incinerators, cylindrical, vertical, horizontal, locomotive, fire tube (straight and bent tube), packaged boilers, flash boilers, scotch marine boilers, re-heat and integral furnace boilers, steam generators (modern central stations), monotube, forced circulation, radiant, mercury boilers—description and sketches.

- **Practical knowledge of various types of steam power plants**: Simple non-condensing, elementary condensing, turbo alternator plants, central stations, high pressure plants, super pressure plants, layout, heat distribution and cycle by description and sketch.

- **Practical knowledge of superheaters, desuperheaters, reheaters, purifiers and attemperators**: Types, locations, advantages, economy, performance, controls, and safety devices.

- **Feed water heaters**: Purpose, types, open, closed and extraction, deaerators, trays, filters, oil separators, back pressure valves, water seals and meters—location, description, sketches, temperatures and pressure carried, saving in fuel, maintenance, malfunctions.

- **Economizers and air pre-heaters**: Types, advantages and disadvantages, locations, soot removal and soot blowers, description and sketches, heat balance diagrams.

- **Feed water conditioning**: Scale, what it is and what causes it; dangers of an excess, impurities generally found in raw waters; characteristics and chemical formulas, methods of conditioning; internal and external treatment, hot and cold lime soda, zeolite, demineralization, ion and anion exchangers and supplementary phosphate treatments, caustic soda and its application, water analysis, parts per million, grains per gallon, tests soap hardness, alkalinity (PM and PH), chloride, sulfite, phosphate, total dissolved solids, use of chelants, value of blow down (both main and continuous), how controlled and measured, meaning of carry over, priming, foaming, embrittlement, corrosion, grooving, how prevented, sketches of one form of external treatment.

- **Boiler construction**: Various forms, plates, rivets, stays, braces, joints, factors, welding, etc. (used under ASME Code), meaning of shearing, tensile stress, modulus of elasticity, bursting pressure, working pressure, factor of safety, meaning of pitch, percentage of plate and rivet strengths, unit, section of plate, how to calculate the strength of boiler heads, shells, stays, efficiency of joints, description and sketches of joints,

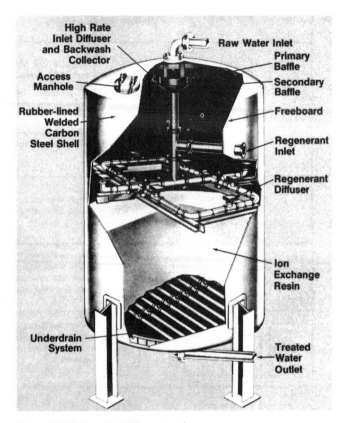

Figure 2-14. Demineralizer vessel.
(Courtesy: Betz Laboratories)

relative strengths of various joints and how calculated, how to
calculate combined efficiency of riveted joints, computation
of working pressure, given tensile strength, factor of safety,
thickness and radius of shell, plus efficiency of joint, how to
stay flat surfaces, including types of stays, allowances for
limitation of stresses, and constants for various types of stays
and doubling plates, how to calculate the area of the heads of
segments of boilers, etc.

- **Boiler settings:** Types, purpose, description and sketches,
  radiation and bridge walls, ignition, mixing and deflection
  arches, baffles, combustion chambers and furnaces, meaning
  of refractory lining, combustion space, dutch ovens, extension

front, importance of sufficient heights over grates, effect of combustion space on the capacity and efficiency, how to build or repair bridge walls and setting, materials and cements used in settings, cyclone primary furnaces, solid refractory, air-cooled or water-cooled walls, water screens, slag tap, dry-bottom and hopper-bottom furnaces.

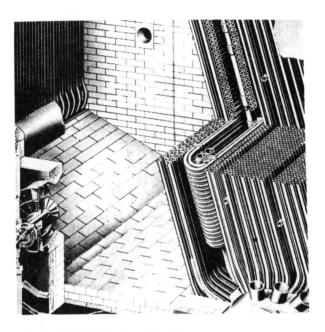

Figure 2-15. Refractory lined furnace.

- **Inspection and testing**: Washing of shell and tubes, cleaning and inspection procedures relative to shell, how often and purpose, cleaning and inspection of setting, auxiliaries, valves and fittings, description of methods used and parts checked, how to inspect, what to look for, possible defects and how corrected, hydrostatic tests and how applied.

- **Mechanical stokers**: Types, overfeed, underfeed, chain grate, vibrating grate, spreader or sprinkler, principles, purpose, description, comparison, sketches, how operated, maintained and repaired, controls, what could go wrong and remedies, feed and driving mechanisms.

- **Oil- and gas-burner installations**: Types of burners, types of oils used in oil-burning installations, heat values, how and where used, methods of burning fuel oil, methods of burning gaseous fuels, dangers connected with oil- and gas-burning installations, description of controls, how to operate and maintain automatic equipment, sketches of industrial burners for gas and oil, sketches of oil-burning-plant layouts.

- **Pulverized coal systems**: Types, storage or central, direct or unit, horizontal and vertical dryers, applications, sketches of pulverized systems, vertical, horizontal, tangential, cyclone, inter-lube burners and feeders, vertical and horizontal pulverizers and dryers.

- **Coal-handling and ash-handling equipment**: Manual, conveyors, scraper flight, bucket, belt apron, grab bucket, telepherage systems, coal crushers and breakers, hoppers, and coal valves, storage internal and external—by pile, silo or bunker, coal-weighing larries, hoppers and coal valves, sketches and descriptions, ash handling—manual, gravity, hydraulic, submerged cross bar, pneumatic, skip hoist and steam jet conveyor, simple and compound cyclone fly ash collectors, sketches and descriptions.

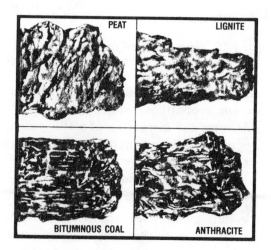

Figure 2-16. Kinds of coal.

- **Rating, efficiency, and testing of steam boilers**: Heat output in steam units, units of evaporation—actual, factor of, and equivalent, evaporation, horsepower rating—ratio of heating surface to performance—(1) efficiency, (2) rate of combustion per grate surface, (3) rate of combustion per pound of fuel, (4) heat transferred per square feet of heating surface, (5) heat liberated per cubic feet of furnace volume—unit testing, comparison with standard or guaranteed results, recording of data and results of a boiler test run.

- **Testing and measuring apparatus**: Coalmeters, fluid meters, steam flow meters and recorders, Bourdon pressure and vacuum gauges, draft gauges, recording draft and vacuum gauges, thermometers, pyrometers, flue gas analyzer apparatus, CO_2 recorders, calorimeters, Ringlemann smoke chart, Bailey smoke density recorder, counters, tachometers, dynamometers, pony brake, boiler control panels—description and sketches.

- **Pumps**: General description of a simplex, duplex, triplex, weir, rotary, centrifugal or vacuum pump, specific knowledge of two types of pumps, knowledge of proper operation and maintenance, where and how installed, lubrication, types of water pistons, packing used and how applied, capacities and applications, dynamic, friction and velocity heads, what could go wrong, how to calculate size of pump for a feed line to boiler at rating or over-rating.

Figure 2-17. Base-mounted centrifugal pump.
(©ITT Corporation — Domestic pump)

- **Piping**: Computation of sizes for water, steam, gas and oil, how installed, allowances for drainage and friction, flow of steam, Napiers, Briggs, Babcocks, Shitzgrass and Fritsoche's rules, sizes of piping for pump discharge and feed line to boiler, pipe fittings, expansion joints, U bends, flanges and gaskets, piping layouts—sketches of sizes, location of piping and valves, pipe hangers, roll anchors and brackets, pipe coverings.

- **Steam engines**: General knowledge of types—simple, compound, triple expansion, quadruple, uniflow, Corliss, high speed and slide valve; specific knowledge of one type, governors, throttle, transformer inertia, centrifugal, cut-off, sketches and descriptions of two types, types of valves, high speed, Corliss, wheelock, simple slide, balanced valves and poppet valves, sketches, how to set valves on various types of engines, methods of lubrication and application to various parts, positive pressure and hydrostatic cylinder lubricators, centrifugal crank pin oilers and chain oiling, how engines are started and how shut down, maintenance and care, dangers, indicators, indicator diagrams and how to interpret the same, indicated horsepower, brake horsepower, pony brake, mechanical efficiency cycle, heat balance of engine.

- **Steam turbines**: Fundamental principles of operation, classification, of steam turbines, general knowledge of types—Impulse, Rateau, Curtiss, DeLaval, Terry, Kerr Sturtevant, Coppus-Reaction, Persons, Allis Chalmers, Westinghouse, Brown Bovery, Ljungstron, General Electric and Worthington, compound, exhaust steam mixed, back pressure, pass out or bleeder turbines; specific knowledge of one type, diagrammatical sketches showing compounding by pressure and velocity in impulse and reaction turbines, assembly of blades, nozzles and diaphragms in various forms, dummy pistons, Kingsbury thrust bearings, etc., turbine glands and gland sealing, hydraulic, carbon ring and labyrinth glands, lubrication forced and combined, automatic regulating valves for auxiliary oil pump, description and sketch, types and properties of oils, sketch of forced lubrication system, oil filters and purifiers, governors and governing, types, descriptions and sketches, emergency stops and controls, descriptions and sketches, a general knowledge of installation, pre-heating and drainage of turbines, starting and stopping turbines, inspection and overhauling, turbine troubles and how corrected, erosion of blading—stresses in turbine rotors, critical speed of turbine rotors.

- **Steam condensers**: Fundamental principles, parallel flow, counter current, low level, barometric, low and high vacuum, siphon or ejector, surface, water- or air-cooled evaporative, air pumps, circulating, dry vacuum, centrifugal entrainment, air ejectors, cooling ponds and towers, types of water, natural and mechanical cooling towers, sketches and descriptions of various types of condensers and auxiliaries.

- **Steam plant auxiliaries**: Super-heaters, economizers, air pre-heaters, feedwater regulators, feedwater heaters, temperature regulating valves, impulse, flow meters, return, float, thermostatic, expansion, dump, differential, siphon, and bucket traps, back pressure non-return, safety, motor operated and reducing valves, boiler feed pump governors, air unloaders, expansion joints, steam separators, calorimeters, oil burners, ejectors, siphons, sight, forced and hydrostatic lubricators, automatic expansion valves, simple steam lock, thermostat and diaphragm valves, meters, steam accumulator, internal drum steam washers or separators—descriptions and sketches.

- **Mathematics**: Knowledge of how to find the area and circumference of a circle and the volume of a cylinder, pyramid and cone, also, the volume of the frustums of pyramids and cones; how to find the area of the segment and zone of a sphere, knowledge of algebra of simultaneous and quadratic equations and logarithms, working knowledge of problems as covered in boiler construction and repair through the use of the A.S.M.E. Code, understanding of the relationship between the sides of right angle triangles, knowledge of the methods of solving problems relating to a train of gears, pulley sizes—rope blocks, screw jacks, levers, wheels, and axles, etc., definite understanding of the meaning of the factor of evaporation, equivalent evaporation, latent heat, total heat; of problems involving the abstraction or addition of heat relative to heating water or making ice or steam, how to solve problems involving heat calculations of steam production, boiler efficiency and horsepower; also to compare boiler performances, how to find the horsepower of an engine from an indicator diagram by a pony brake or a dynamometer, how to solve problems involving specific gravity and specific heat, Boyles and Charles Laws, the parallelogram of forces, Newton's Law, how to solve questions covering economy of operation relative to fuel and labor costs, how to calculate heating surface and horsepower ratings of various types of boilers.

- **Heating, air conditioning and ventilation:** Fundamentals of control and measurement, electric and pneumatic control circuits, zone and unit controls, controllers, actuators, types, valves and relays, piping, fans and duct work, vacuum and plenum systems in mechanical ventilation, radiation, gravity and vacuum steam systems, designs, installations, operation and maintenance, descriptions and sketches.

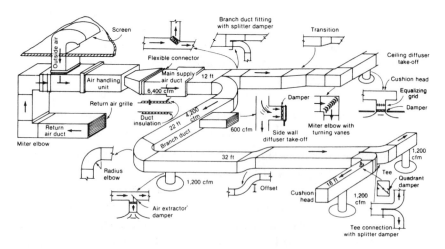

Figure 2-18. Low-velocity air duct system.
(Courtesy: Building Owners and Managers Institute)

- **Refrigeration:** Cycle of compressor system, theory of how heat is abstracted, common refrigerants and their physical properties, direct and indirect systems, purpose, location, description and sketches of a safety head, cylinder, valves, safety valves, condensers, receiver, compressor, oil separator, evaporator, flow equipment such as automatic purgers, dehydrators and heat exchangers, types, descriptions and sketches of condensers, of hand, automatic and thermostatic expansion valves, low and high side float valves, types and descriptions of compressors, single and double acting, rotary, centrifugal and steam jet, types and descriptions of evaporators, flooded, dry expansion, convection and forced draft, electrical controls and control valves, back pressure and temperature controls, constant pressure, snap action, check water regulating, valves, high pressure and low water cut outs, pressure relief devices, lubrication, selection and characteristics of oils, pour point, viscosity, cloud point and moisture content, splash feed and forced feed, adding or changing oil, effects of too much oil, compressor drives, electric, steam diesel or gas engine, operating and maintenance of compressor system, starting and stopping, how to charge, reduce or remove refrigerant from system, pumping out condenser and defrosting, dangers in operating, effect of air and how controlled, purging; how ammonia leak is found and stopped, annual inspection and overhaul, compressor calculations, cylinder displacement, volumetric efficiency, power requirements for condenser, brine in refrigeration, chemistry, brine pumps, corrosion control, congealing, hold over tanks and eutectic plates, liquid cooling and ice making, absorption systems—intermittent, continuous and Electrolux, description, application, cycle and sketch, the Refrigeration Code.

- **Air and gas compressor systems**: General knowledge of various types of compressors, efficiency, isothermal and adiabatic compression, air compressor valves, automatic, mechanical types, descriptions, sketches, maintenance, speed regulation, governors on steam or diesel engines, constant speed, unloaders, closed and open intake, discharge, clearance pocket, description and sketches, intercooler, aftercooler, receivers, cooling systems, descriptions and sketches, lubrication, types of oil and how applied, danger of too much oil, safety devices, inspection and overhauls, dangers in operation—causes of explosions, internal combustion engines—principles, operating and maintenance, installation and general layout of air compression systems, gas compression—natural, illuminating, gas absorption process, vapor recovery, stabilization and refractionation, application of compressed air, free air, standard air, heat of compression, volumetric, compression, mechanical and overall efficiencies, dew point and humidity.

• **Electricity**: A practical knowledge of electromagnetism, ohms, Coulomb, Joules, and Kirchoffs Laws, laws of induction, self-inductance, mutual inductance, current in revolving loop, direct current generators, field frame, magnetic poles and windings, armature core and windings, commutator brush rigging and brushes, sine curve, commutation, excitation, characteristics of series, shunt and compound generators, operation of machines in parallel, starting and control of D.C. generators, direct current motors, types characteristics and applications, effect of armature reaction, how to install brush rigging and brushes of controls and connection diagrams, no voltage release, overload release, alternating current generators, alternators, types, single and polyphase, revolving armature, revolving field induction, self excited, separately excited, slow or high speed, how does A.C. produce a revolving magnetic field, description of mechanical construction, rotor, stator, slip rings, bush gear, damper or amortisseur windings, exciter—how started, how controlled, how stopped, phasing out, synchronizing, operation in parallel, division of load, starting, pull in and pull out torques, description of starting equipment and connection diagrams, voltage regulation and regulators, A.C. motors, theory, revolving magnetic field, slip, running speed, no load and full load, types, squirrel cage, wound rotor, single phase, synchronous motors, converters, construction, characteristics and applications, multispeed squirrel cage motors, classification of A.C. motors, open, semi-enclosed, totally enclosed, fan cooled, enclosed ventilation, vertical motors, starting and control of A.C. motors, types of controls and connection diagrams, no voltage released, overload release, switch gear, circuit breakers, relays, rheostats, switchboards, instrument transformers, bus bars, disconnects, description and sketches, connection diagrams for switchboards and motor converter sets, electrical meters—ohmmeter, volt-meter, a ammeter, wattmeter, watt hour meter, frequency meter, synchroscope, fundamentals and applications, power factor—meaning, factors controlling power factor, effect on generators, motors and lines, how improved, transformers—principles and types, core and shell types, construction of core and coils, cooling, windings and connections, losses and regulation, electrical calculations and conversion factors, horsepower ratings and speeds of motors and generators, comparison with boiler and engine horsepower, ratings of circuit breakers, computation of size of motor, generator or transformer for various applications, service factor, calculations need in power factor improvement, general knowledge of electric boilers— construction, controls, operation, lubrication, oil ring, grease closures for all and roller bearings, inserters and batteries.

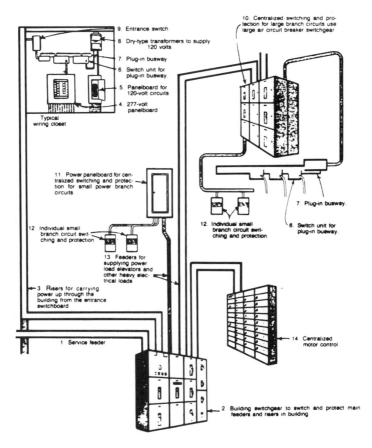

Figure 2-19. Building electrical system.
(Courtesy: Building Owners and Managers Institute)

- **Lubrication**: Oils—vegetable, animal and mineral, physical tests and applications, lubrications, oil cups, telescopic oil ring, centrifugal, pendulum, hydrostatic, sight feed, forced feed, gravity, compressed air feed, filters and by-pass systems.

- **Physics, mechanics and chemistry**: Matter, force weight, two or three acting at a point, triangle, polygon and parallelogram of force, movements, parallel forces, couples, center of gravity, work, mechanical advantage, velocity ratio, energy, power, efficiency, friction, velocity, acceleration, inertia, transmission of motion and power hydraulics, machines, levers, pulleys, wheel and axle, inclined plane, heat, measurement, expansion rate, Charles-Guy Lussac, Boyles Laws, theory of siphon and barometer, losses in power and heat transmission through modern steam plant, Joules equivalent, chemistry, quantitative laws, colloids, periodic law, structure of atom, atomic weights of elements related to combustion, a general knowledge of such metals as copper, brass, iron, magnesium, calcium, zinc, aluminum, lead, antimony, nickel, steel, the alkali metals, lithium, sodium, potassium, etc.

- **Working sketches, as required of general equipment in the above sections**: Need not be to scale, but should be in general proportion, details of parts could be enlarged adjacent to sketch if necessary.

- **Machine and mechanical drawing of a test piece**: Showing as many views as will show all dimensions, drawing must be to scale, neatness and accuracy are essential, details must be shown sufficiently that an identical piece can be made from the drawing.

- **Erection**: Foundations, layout, form building, templets, lining up engines, method of bedding shaft, fitting, testing alignment, grouting, assembly, adjustment, testing, etc.

- **Management**: The human factor, methods, time and motion, work analysis, work improvement, records, charts and diagrams, work sampling, quality control.

- **Economics**: Elementary economic concepts, evolution of economic activity, economic systems, production, organization, marketing, risk transportation, consumption, supply and demand, money credit and banking, changing value of money, distribution of income, population problems, labor problems, public utilities.

- **Ecology**: Ecosystems, water quality control, general criteria for all waters, water treatment, waste water treatment, sterilization, air quality control, impurities in the air, sources of air impurities, particulate matter.

- **Maintenance programs**: Meaning and scope of maintenance control, developing yardsticks and benchmarks for measuring maintenance principles of control, forecasting, preventive maintenance programs, work authorization, planning, routing, material control, tool control, scheduling.

- **Real property improvement**: Nature and scope, determining the value of property and equipment, inventory, depreciation, estimating endurance and use, control of repair and rehabilitation.

- **Foremanship/Supervision**: Human relations, the art of supervising, controlling manpower, establishing controls, work procedures, establishing training, safety, work schedules, establishing control through reports and follow-up, OSHA instructions.

- **Technical writing**: The business letter, memoranda, the data sheet, writing technical explanations, report writing, gathering data, format, use of different types of reports, reading of related subject matter.

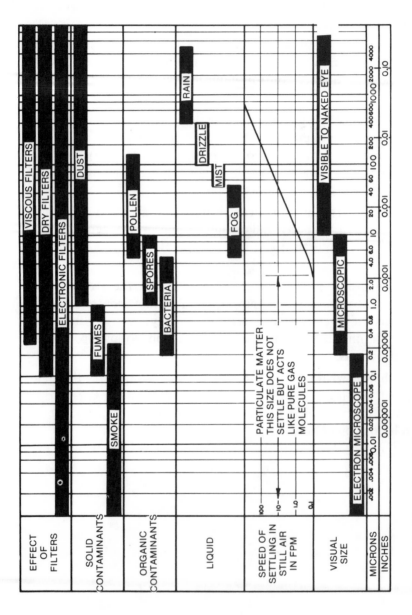

Figure 2-20. Sizes of particulate matter.
(Courtesy: Building Owners and Managers Institute)

Chapter 3

A TYPICAL WORKDAY

Before we delve too deeply into the content of this chapter, I suppose I should qualify the title I've chosen for it; otherwise my brother engineers will argue, vehemently, that no such animal roams the Stationary Engineering jungle. And I can't disagree with them. My own experience has been that each new day on the job differed from the one preceding it. Like fingerprints and DNA maps, no two are ever the same. I guess that's why I've never lost interest in the profession. But for every difference that exists, there is a similarity and it's that common ground we'll be exploring here.

SETTING THE STAGE

Considering the magnitude of the responsibility an operating engineer accepts during the course of his shift, it stands to reason he should be mindful of what's happening in the plant beforehand. How do you go about acquiring this knowledge, you ask? How about this scenerio? We've travelled down time's bumpy road; you just received your First Class license . . . (congratulations!) . . . I'm about to retire and since you'll be my permanent replacement, the company has allowed us one week to work together before turning you loose on your own. Welcome to the real world. Officially, our shift begins at 3 p.m. and ends at 11 p.m.—a straight 8 hours and out the door. Now that schedule may suit company policy, the union agreement and the guy putting his time in strictly to collect a paycheck, but we are professionals and our job isn't governed by the ticks of a clock. Meet me in the parking lot at half-past two.

ARRIVING AT THE PLANT

Hi guy! I'm glad you made it on time; we've got a lot to do before going inside. Let's begin by taking a tour of the grounds to check on the building structure and exterior mechanicals. Here are some things to be mindful of which may indicate a present problem, a past failure or a future concern:

- cracks in or masonry missing from the power house stacks
- excessive smoke issuing from the stacks
- missing or plugged vent caps
- foam cascading from the cooling tower
- smoke or noise issuing from the cooling tower
- steam billowing from safety valve discharge pipes
- blockage of air intakes by leaves, trash or snow
- proper adjustment of louvers
- pooling of water around drains
- wet, soppy ground by the building foundation
- cracked and buckled concrete drives and walkways
- evidence of leakage around buried fuel tanks
- overheated or noisy power transformers
- malfunctioning outdoor sensing devices
- blinking signs or lights
- cleanliness of photo-electric eyes
- proper operation of automatic doors
- broken windows and door locks
- blocked openings on outdoor condensers
- burned exterior surfaces on incinerators
- frayed or fallen power lines
- large accumulations of trash
- improperly stored combustibles
- noisy exhaust fans
- leaking fire hydrants
- evidence of damage to buried utilities

TAKING OVER THE WATCH

Okay, we took a good look around outside and found nothing of consequence to report or take action on. I hope our luck holds out. Let's go inside and find out what the first-shift man has to say before we relieve him. On our way in we'll check all the main shut-offs to make certain they're correctly positioned. Let's go and see George.

This is the engineer's office, but George isn't here. He must be out making his final rounds. While we're waiting for him to return, call the weather service . . . here's the number. Jot down what precipitation, humidity and temperatures they expect for this evening. Why call the weather service? For two reasons. First, you'll need the information to make equipment adjustments later on during your shift. More importantly, you now know that the telephone works. In the event of an emergency, this telephone is your only contact with the outside world. Potentially, it represents the difference between life and death or a damaged piece of equipment and the demolition of a multi-million-dollar plant. Who do you call for what? Good question. The roster is posted on the bulletin board over the desk. As you can see it's quite lengthy. It's comprised of personnel and/or organizations that are accessible 24 hours a day and contains the telephone numbers of:

- selected members of the corporation
- local fire and police departments
- the county emergency management officer
- the hazardous material squad
- local hospitals and paramedics
- electric power and gas companies
- fuel and water companies
- equipment manufacturers' service departments
- the fire warning system monitoring center
- bulk cylinder gas companies
- the sewage lift station
- contracted security firm offices

Of course when a situation manifests itself, you must respond to it based on your judgement of the seriousness of the problem and its potential for becoming worse. I believe in the old adage "better safe than sorry," but don't be too quick to call these organizations for help at the first sign of trouble. In many instances, your call may mobilize more manpower than was used to build the pyramids. Just kidding, of course, but it will seem that way if your call turns out to be a false alarm. And you've heard of the boy who cried wolf, I presume? When confronted with a problem, if you have the time, call one of your brother engineers for their input first. Their names are listed here too, and they'll be more than happy to help you out of a predicament they ve probably found themselves in at one time or another.

How's that for timing. Here comes George now. He knows about your starting tonight and I've arranged to have him spend a little time with you to get acquainted and answer any questions you may have regarding his shift activities. I'm going down to the cafeteria to get us a cold drink. Introduce yourself and while you're reviewing the engineering log cover these areas with him:

- the general condition of the plant
- which pieces of equipment are on line
- broken or missing system components
- malfunctioning equipment or instrumentation
- results of water testing
- chemical treatments administered
- failed tests of safety devices
- special procedures performed
- abnormal pressures or temperatures
- nuisance tripping conditions
- unusual occurrences of any kind

I see George has gone home. Here's a pitcher of iced tea; they were all out of canned drinks. How about grabbing a couple of those cups off of the shelf . . . that's right, the new one is yours. The guys pitched in to buy it for you. Well, how did you and George make out? Fantastic! George is a good man; you'll enjoy working with him. By the way, he's the one that painted your name on your cup. What did he have to say about his operations? Uh-huh. Un-huh. That's great! It appears that the guys took care of everything for us to make your first day an easy one. Okay, sign us into the log and let's get started.

MAKING ROUNDS

To their detriment, some companies prescribe rigid routines for monitoring equipment operations on their premises. All that is accomplished by such archaic standards is the completion of forms which get tucked neatly away in file cabinets, never again to see the light of day. Enlightened management realizes the futility of such a program in this type of operation and allows the engineer a more flexible schedule. I personally take a two-fold approach to the problem, making one set of rounds to assure the proper functioning of the plant's mechanicals and a second set of rounds to record pressures and temperatures . . . etc. You'll see the difference in the two as we proceed through them. We'll start with the mechanical rounds first. The idea behind the initial set of rounds is to make a complete pass through the machinery spaces without being encumbered by the need to make repairs, perform tests or record readings. If properly done, at tour's end, we should have a good feel for the operation and be better able to plan out the remainder of our shift. We're about to embark on a very sensual trip through the plant. So open your eyes, and put your ears on. Here's our itinerary for the journey.

- visually survey the general condition of the plant
- listen for any strange or unusual sounds
- feel the equipment for evidence of overheating
- observe fuel oil tanks for evidence of leakage
- sound fuel and water tanks to determine proper levels
- feel the equipment to determine excessive vibration
- look for burned out ceiling lights in machinery spaces
- note where indicator bulbs are burned out or missing
- check boiler stacks for proper operating temperatures
- try by-passes and alternate units
- check the water level in the sump and trial the pump
- observe machinery for loose nuts, bolts and mounts
- note any discoloration of exterior metal surfaces
- take notice of abnormal temperatures or pressures
- check for proper water levels in boiler gauge glasses
- watch for evidence of steam, water, fuel or gas leaks
- draw water samples from boilers and cooling towers

- discharge any water that has accumulated in the air tanks
- check oil levels in all rotating and reciprocating equipment
- ascertain that the boiler's low-water fuel cut-off works
- operate soot blowers through their entire range of motion
- make certain all safety accoutrements are accessible
- lift safety and relief valves by hand
- make sure fire extinguishers are in place and fully charged
- operate eye wash stations and decontamination showers
- listen to steam traps for evidence of blowing through
- check evaporators for excessive accumulations of ice
- watch out for loose deck gratings and pipe hangers
- be alert for short cycling and extended machinery operation
- note if operating certificates and licenses have expired
- check the stock room for adequate levels of consumables
- observe packing glands for proper drip
- make certain emergency backup lighting systems work
- check battery electrolyte levels and charging equipment
- ensure that the plant has hot and cold running water
- witness that the ventilation system is working properly
- listen for noisy bearings and gear assemblies
- make sure that all dampers are operational
- inspect valve bodies, vessels and piping for cracks
- check the furnace interior for spalled or missing refractory
- observe the color of the flame in the combustion chamber
- look for tagged or locked-out valves and switches
- keep alert for evidence of corrosion
- make sure all shafts move freely
- note broken or missing thermometers or gauges
- replace missing cover plates and machine guards
- look for broken glass and oil spills
- check expansion tanks for waterlogging
- listen to and feel stationary electrical apparatus
- observe refrigerant flow in sight glasses
- bleed air from all closed water loops
- check incoming water and gas pressures

- note the value of the incoming primary voltage
- watch for leaking sprinkler heads
- check the central clock station for correct time
- look for broken couplers on idle equipment
- visually inventory your chemical supplies
- watch for loose belts and/or poor alignments
- feel heat exchangers for temperature differences
- check manometers across air filter bank

Here we are back at the starting gate. You can set those water samples on the counter; they have to cool before we run our tests on them. I enjoyed our little sojourn in the power plant; was it good for you? That's nice; I was hoping you'd get something out of it. How's that? No, thanks, I gave them up last November, but feel free to light one up tor yourself, if you like. Now that we're away from the commotion of the machinery bays, do you have any questions? Yes, it's true; that was an extensive checklist, but by no means does it cover every contingency you'll encounter. As you become better oriented to the plant and more familiar with its operational nuances, you'll develop your own routines; in the meanwhile these guidelines should serve to keep you out of trouble. No, that's not all there is to it! Granted, you're a quick study, but you've got a way to go before I put my stamp of approval on you.

Remember, the first tour gives you a grasp of the plant's operating status, enabling you to make an informed decision as to how you will proceed for the remaining 7 hours of your shift.

SCHEDULING THE WORK

Let's review our findings and plan a course of corrective action, bearing in mind of course, that we have readings to take and operator checks to make. Aren't they the same thing, you ask? Oh, boy; I can see this is going to be a long night. Listen up! Operator checks deal with the how-to of power plant operations. They are the application of the operator's expertise in the performance of his job, involving the alternation, testing, calibration, care and storage of equipment, system components, control, regulating and monitoring devices and instrumentation. Readings are simply the documentation of a device's physical condition at the moment of observation. It's apparent from your question, my friend, that a more thorough study of operating procedures may be in order. What do you say we cover them in another chapter, otherwise we're never going to finish this one? Great! Thanks, guy. With that out of the way, give me the discrepancy list we compiled during our mechanical rounds. I'll use it to prepare our schedule for the rest of the evening. While I'm doing that, why don't you go down to the cafeteria before they close and bring us back a pot of coffee? Huh? No, no sugar for me, but a couple of creamers if they have them.

Welcome back. Thanks for running after the Joe. Here's our game plan. Look it over while I pour us a cup. If we don't run into any snags, we have a relatively easy night ahead of us.

4:00   Repair coupler on #2 feed pump. Check sump pump linkage for hang-up.

4:30   Make rounds to take readings from the equipment. Replace burned out indicator lamps on boiler #1 and chiller #1 control panels. Blow down float controls.

5:30   Analyze water samples taken from the cooling tower and boilers . . . adjust the bleed off, blowdown and treat waters based on results. Test water softeners for hardness and regenerate if necessary.

6:15   Adjust the belts on air-handling unit #3.

6:30   LUNCH

7:00   Perform flue gas analysis and adjust burner controls as needed.

7:30   Perform slow drain test of low-water fuel cut-off controls.
       Purge chillers. Check exhaust fans on roof.

8:00   Take second set of equipment readings.

9:00   Replace burned-out ceiling lights in boiler room and cafeteria. Advance roll filters in air handler #1.
       Replenish chemicals from stock.

10:00 Make final mechanical check and log entries.

CALLING IT A DAY

I'll be damned . . . that's the first time in my 25 years as an operator that everything fell into place . . . no interruptions, breakdowns or telephone calls. As a matter of fact, we could have used our game plan to make the log entries. Slap me; I must be hallucinating. It's still early, but I see Gary is here to start the graveyard shift. I'll introduce you two, then I want you to take him through the same routine you and George went through at the beginning of our shift.

What? Oh, sure; I think you did a super job for your first night out. Yes, of course I believe you'll be able to handle it by yourself . . . eventually. Certainly you can . . . look, don't push your luck; the truth is you fit in real well here, but you're still a little green. Just drive tough and I'll help you smooth out your ride. Tomorrow is shaping up to be a slow day. If you like, we can take some time to go over those operator checks we talked about. Uh-huh. Now you're catching on. And a good-night to you too, sir.

Chapter 4

SELECTED
OPERATING PROCEDURES

Good afternoon. I'm happy to see your first night on the job didn't scare you away. You say you're ready for some intense one-on-one? Hold on a second, friend; as I recall yesterday's conversation, we agreed that tonight you'd be jumping through hoops, not stuffing them. I know, I know, but you really can't afford to play around at this stage of the game; remember, in 4 days you'll be on your own. We've got a lot of territory to cover, so why don't we get our initial rounds out of the way and get on with our game plan. Of course, if you feel like I'm pressing you, you can always call foul . . . My goal is just to make a couple of points . . . okay, okay, I apologize . . . so take me to court.

Yesterday we touched on your need for additional tutoring in the area of operator checks. As the number of operator's procedures available for discussion are probably comparable to the average number of baskets made in a roundball tournament, I've decided that we would accomplish more by having you bounce your questions off of me than if I were to shoot a lot of information at you that you're already aware of. Actually, if you're game, I can walk you through some of the procedures while we're doing mechanical rounds. The ball is in your court. You pick the topic and I'll checklist a procedure for you.

HOW TO . . .

. . . Start Up A Steam Boiler

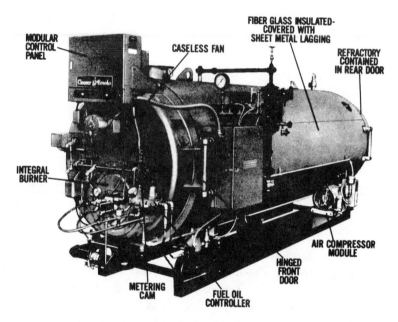

Figure 4-1. Steam Boiler.
(Courtesy of Cleaver Brooks)

1. Determine the status of other units in the battery
2. Thoroughly acquaint yourself with the equipment
3. Read the manufacturer's instruction manual
4. Visually inspect the unit for obvious problems
5. Check all instruments for proper calibration
6. Move all the control switches to the OFF position
7. Make sure there is adequate combustion air
8. Check valves for correct positioning
9. Test electrical power supplies for proper values
10. Dry run control devices and interlocks
11. Check the fuel and feedwater supply tranes
12. Isolate the boiler from the steam header

13. Open the air vent on top of the shell
14. Admit feedwater until it registers in the gauge glass
15. Purge the furnace, passes and breeching with air
16. Check for proper settings of all operating controls
17. Manually reset the operating controls
18. Clean the fire-eye and ignite the main burner
19. Observe burner operation through two cycles
20. Visually inspect the exterior of the boiler
21. Close the air vent when steam issues from it
22. At 75% of set pressure, lift safety valves by hand
23. When unit nears header pressure, open the main

. . . Boil Out Oil Contamination

1. Adhere to the manufacturer's chemical and pressure guidelines
2. Remove all foreign objects from the unit's interior
3. Isolate the unit from the header
4. Open the air vent at the top of the boiler
5. Admit hot water to the normal operating level
6. Pour in the prescribed mixture of compound
7. Button up the boiler and continue adding water
8. Reset operating limits to one-third of normal
9. Light off the burner and maintain on low fire
10. Close the air vent when steam issues from it
11. Boil solution for a minimum of 5 hours
12. Shut the unit down and open the air vent
13. Allow the boiler to cool naturally
14. Ask the water company for permission to dump
15. Drain the entire solution to a safe point of discharge
16. Unbutton all inspection and access ports
17. Flush all internal surfaces with a high-pressure hose
18. Replace the manhole and handhole gaskets
19. Install new gauge glasses and packing
20. Admit hot water to the normal operating level
21. Treat the feedwater per chemical company parameters

22. Reset the control limits and fire-off the burner
23. Bring the boiler up to the header pressure
24. Open the main steam valve and resume normal operation

. . . Clean A Boiler Gauge Glass During Operation

1. Pour some household ammonia into a cup
2. While steaming, close the top and bottom gauge glass valves
3. Open the gauge glass drain line valve
4. Crack the top valve to blow the water out of the glass
5. Close the top valve to form a vacuum in the glass
6. Place the drain line tip into the ammonia
7. Let the vacuum draw the ammonia into the glass
8. Crack the top valve to blow the ammonia out of the glass
9. Repeat the process until the glass is clean
10. When the glass is clean, clear it and close the drain valve
11. Open both gauge glass valves and resume normal operation

. . . Blowdown A Water Column

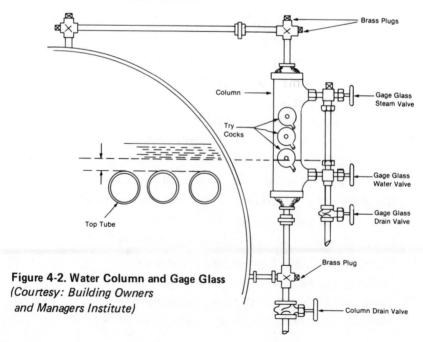

Figure 4-2. Water Column and Gage Glass
(Courtesy: Building Owners
and Managers Institute)

1. Close the top valves on the water column and gauge glass
2. Blow water through the gauge glass
3. Blow water through the water column
4. Close the bottom valves on the water column and gauge glass
5. Open the top valves on the water volumn and gauge glass
6. Blow steam through the gauge glass
7. Blow steam through the water column
8. Open the botton valves on the water column and gauge glass
9. Resume normal operations

... *Replace A Broken Gauge Glass*

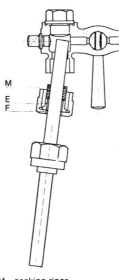

Figure 4-3. Water Gauge Glass
(Courtesy: Building Owners and Mangers Institute)

M = packing rings
E = ferrule
F = packing nut

1. Close the steam inlet valve then recrack it open
2. Close the water inlet valve
3. Open the drain line and reclose the steam inlet
4. Monitor the water level using the try cocks
5. Remove the rods protecting the gauge glass
6. Back off the upper and lower retaining nuts
7. Remove the broken gauge glass and seals

8. Cover the bottom opening with a gloved hand
9. Flush broken glass by opening the water inlet valve
10. Quickly close the water inlet valve
11. Insert the new glass and seals
12. Tighten the retaining nuts by hand
13. Snug up the nuts a quarter turn with a wrench
14. Replace the gauge glass guard rods
15. Close the drain line valve and crack open the steam inlet
16. After the glass warms open the steam inlet fully
17. Fully open the water inlet valve
18. Blow down the gauge glass and return it to service

. . . Store A Boiler — Short Term

1. Consult the operator's manual for instructions
2. Shut down the boiler and let the gauge pressure drop to zero
3. Close the main steam discharge valves
4. Purge the furnace of unburned fuel vapors
5. Open the air vent at the top of the shell
6. Drain the boiler through the blowdown line
7. Clean and inspect the watersides and firesides
8. Button up the unit and begin filling slowly with deaerated water
9. Condition the feedwater as prescribed by your treatment consultant
10. Fill until flooded with water well above ambient temperature
11. Cut back the operating control limits to one-third of normal
12. Bring the water up to steaming temperature on low fire
13. Secure the burners after the entrained oxygen has been driven off
14. Close the air vent and check the boiler connections for leaks
15. Maintain water pressure in the unit above atmosphere
16. Conduct weekly water tests and add treatment as needed
17. Circulate the water for 1 hour each time treatment is added
18. Reduce chemical levels before returning the unit to service

19. Reset control limits to their original settings
20. Place a "WET STORAGE" sign on the burner assembly

. . . Store A Boiler — Extended Term

1. Consult the operator's manual for instructions
2. Shut down the boiler and let the gauge pressure drop to zero
3. Close fuel valves and disconnect burner fuel lines
4. Purge the furnace of unburned fuel vapors
5. Close the main steam discharge valves
6. Open the air vent at the top of the shell
7. Allow the boiler to cool naturally
8. Drain the boiler through the blowdown line
9. Open all inspection ports and access openings
10. Clean and inspect the watersides and firesides
11. Repair and replace missing and broken parts
12. Completely dry all boiler surfaces
13. Place trays of moisture absorber in the waterside section
14. Cap off or block all lines leading to the unit
15. Replace all gaskets and reseal the unit
16. Coat the fireside surfaces with a rust inhibitor
17. Circulate warm dry air through firesides
18. Install the stack cover if applicable
19. Inspect occasionally and replace dessicant as needed
20. Place a "DRY STORAGE" sign on the burner assembly

. . . Ensure Steam Trap Performance

1. Locate your plant's traps on a single line drawing
2. Identify their types and operating characteristics
3. Develop a maintenance and performance documenting program
4. Establish testing based on manufacturer's recommendations
5. Observe trap operations using established examining techniques
6. Rate the operating efficiency of individual traps
7. Analyze the data to determine if recurring problems exist
8. Record the findings for future comparison
9. Schedule traps for repair or replacement

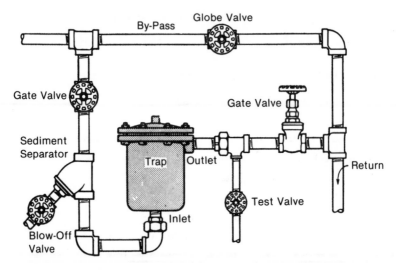

Figure 4-4. Typical Pipng Arrangement for Inverter Open-Float Steam Trap
(Courtesy: Building Owners and Managers Institute)

. . . Install A Battery Bank

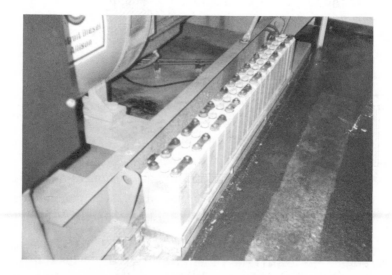

Figure 4-5. Battery Bank

1. Choose a location out of the normal way of travel
2. Condition the space to be cool, dry and well ventilated
3. Assemble the battery rack per manufacturer's instructions
4. Ensure eye and body wash stations are functional
5. Don skin and eye protection gear
6. Extinguish all open flames in the area
7. Unpack and inspect the batteries for leakage
8. Position the cells in their permanent slots
9. Make cell terminal connections
10. Follow guidelines for filling and changing
11. Take and record hydrometer readings
12. Connect the automatic charger if applicable

. . . Pack A Stuffing Box

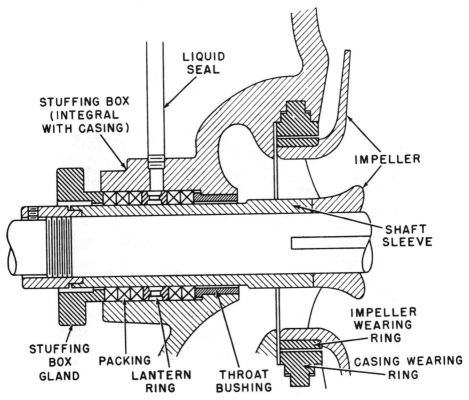

Figure 4-6. Centrifugal Pump Stuffing Box

1. Shut down and lock out the unit to be repacked
2. Completely remove all the old packing
3. Determine the number of rings needed
4. Inspect the shaft for evidence of scoring
5. Make sure the new packing is the right size and type
6. Cut new rings to fit the shaft
7. Dip the rings in oil and insert each separately
8. Draw each ring up with the gland nut until filled
9. Snug up on the packing gland nut to assure seating
10. Back off the gland nut and resecure finger tight
11. Restore the unit to normal operation
12. Check for proper drip rate

. . . Degrease A Fuel Oil Heater

1. Provide for safe storage or disposal of all purged fluids
2. Extinguish all open flames and ventilate the area
3. Totally isolate the unit from the fuel oil system
4. Vent and thoroughly drain the fuel oil heater
5. Fill the unit with low-pressure steam and continue to drain
6. Circulate a degreasing solution per manufacturer's guidelines
7. Drain and blow the unit out with low-pressure air
8. Rinse the interior thoroughly with hot air
9. Refill the unit with oil and return to service

. . . Perform A Refrigerant Leak Test

1. Use a properly calibrated electronic leak detector or halide torch
2. Ventilate the area is which the leak test is to be done
3. Eliminate as much surrounding noise as possible
4. Listen for hissing around common leak locations
5. Look for obvious signs of oil leakage
6. Perform a soap test on all connections
7. Tighten all fittings and connections
8. Performance test the detector using a refrigerant sample

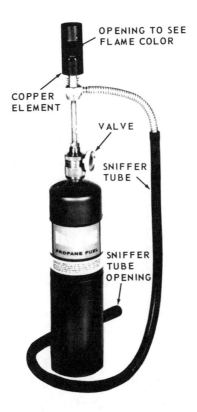

Figure 4-7. Halide Torch
(Bernz O Matic Corp.)

9. Survey the system starting at its highest point
10. Probe each component thoroughly
11. Be alert for evidence of rubbing and misalignment
12. Check all brazed or soldered joints and seals
13. Remove insulation around suspected areas
14. Mark all areas where leaks are detected
15. Repair leaks and resume normal operations

. . . Evaluate A Refrigeration System

Figure 4-8. Refrigeration Evacuation Pump

1. Choose an appropriately sized vacuum pump
2. Discharge the system refrigerant to the receiver tank
3. Replace all the dryers in the system
4. Isolate the system from all mechanicals
5. Tighten down all system valve packing glands
6. Provide the pump with new vacuum pump oil
7. Connect the pump using rigid metal conduits
8. Evacuate the high and low sides to 20 microns
9. Continue pumping for several hours
10. Frequently check the quality and level of the pump oil
11. Close the pump inlet valve and stop the pump
12. Let the system stand overnight
13. Monitor the system for vacuum loss
14. Recharge the system and resume operation

... Operate An Electric Arc Welder

Figure 4-9. Electric Arc Welder

1. Visually inspect all welding components
2. Make certain a properly sized power receptable is available
3. Use the welder only in well-ventilated areas
4. Position the welder out of the way of travel
5. Select the electrode diameter using the material thickness gauge
6. Match the electrode metal thickness to the size chart
7. Match the stencilling on the rod flux to the welder band
8. Don a welder's helmet and gloves
9. Lock in the heat selector dial to rod diameter value
10. Connect the work clamp to the work
11. Turn the switch to the ON position
12. Adjust the amperage output as needed
13. Turn the switch off and allow the work to cool
14. Unplug the welder and unclamp the work
15. Properly store the welder and its accessories

Chapter 5

BASIC
ENGINEERING CONCEPTS

You appeared a tad hesitant carrying out the instructional checklists we just reviewed. Don't worry about making a mistake or two; I'm here to help you through them. Besides, I absolutely refuse to allow people to make fools of themselves in my power plant. How's that? Sure; I'll be glad to run a little theory by you. Even us oldtimers need a back-to-basics session now and then. A short review may do me some good too. Tell me where you'd like to start and I'll do my best to fill you in. The beginning? Uh . . . right. What do you say I just choose a few topics that I feel you might most benefit from discussing? Great.

PRESSURE AND VACUUM RELATIONSHIPS

It's been said that the most pressure-packed position which a person can hold and still be in a vacuum is that of president; whether of a company or country. Stationary Engineers are living proof that nothing is further from the truth. They work with extremes of pressures and vacuum each day and the only board meetings they attend are spelled b-o-r-e-d. Figure 5-1 shows the relationship between the kinds of pressures to which power plants are subjected.

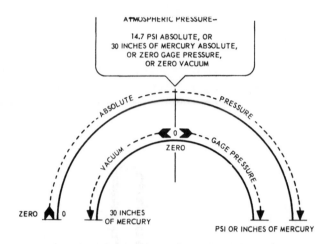

Figure 5-1. Pressure-Vacuum Relationship

It appears that not all pressures were created equal. Actually there aren't different kinds of pressure, just different ways that pressure is represented. Let's review the topic and you'll see what I mean.

Atmospheric Pressure — the weight of all the air that makes up the atmosphere surrounding our planet is brought to bear on its surface as a consequence of the earth's gravitational pull. Atmospheric pressure is a measurement of the weight of the air over a given area (usually a square inch) taken at some point in a column of air which extends from the earth's surface to an indefinite point in outer space. This is generally accepted as 14.7 pounds per square inch (psi) at sea level. In other words, each column of air having a 1-inch-square base which extends from the earth to space contains approximately 15 pounds of air by weight. If you measured 1 square foot of area, the total weight increases to over 2,000 pounds or approximately a ton of air by weight, since there are 144 square inches in a square foot and each of them contains a column of air weighing, for the purpose of simplifying the calculation, 15 pounds.

i.e., 15 (pounds of air) x 144 (square inches) = 2160 (pounds of air)

If this is true (would I lie to you?), then it stands to reason that the weight (pressure) of the air would be less at altitudes above sea level and more on areas located below sea level. This accounts for the facts that the air is "thinner" in the mountains and that water boils at temperatures increasingly lower than 212° Fahrenheit as you ascend them. The device used for measuring atmospheric pressure is the barometer.

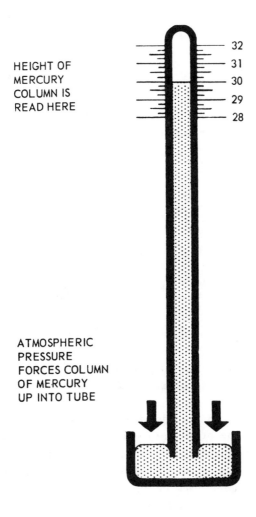

HEIGHT OF
MERCURY
COLUMN IS
READ HERE

32
31
30
29
28

ATMOSPHERIC
PRESSURE
FORCES COLUMN
OF MERCURY
UP INTO TUBE

Figure 5-2. Operating Principle of Mercurial Barometer

Gage Pressure — given a pressure gage which is graduated from zero and read in pounds per square inch (psi), gage pressure is the pressure which is indicated by the gage dial at or above atmospheric pressure. A dial pointing to zero indicates that the pressure being measured is exactly the same as the pressure of the surround-atmosphere. A dial fixed on 100 psi indicates that the pressure being measured is 100 psi in excess of atmospheric pressure. A dial pointing below the zero mark indicates that the gage is broken or that the pressure being measured is below atmospheric (in the vacuum range) and that a problem may exist within your pressure vessel or that an inappropriate gage is being used to measure the pressure within it.

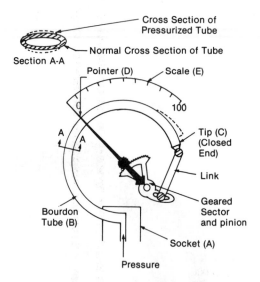

Figure 5-3. Bourdon Tube Gage
(Courtesy: Building Owners and Managers Institute)

Vacuum — when the pressure existing in a vessel is below atmospheric, the vessel is said to be in a vacuum. Actually, pressure still exists within the vessel but at a reduced level. Vacuum gages or the vacuum side of compound gages are used to measure this condition. Just as pressure gages measure pressures at or above atmospheric in psi, vacuum gages measure pressures below atmospheric in inches of vacuum. Two inches of vacuum are approximately equivalent to 1 psi. So, for instance, if you should read a vacuum gage and it shows a vacuum of 16 inches, that would indicate that the pressure you were measuring was 8 psi less than the surrounding atmosphere or approximately 7 psi at sea level.

Ex: 16 (gage reading in inches) ÷ 2 (inches per psi) =
 8 psi (pressure below atmospheric)

 15 psi (atmospheric pressure) − 8 psi (pressure below atmospheric) =
 7 psi (pressure remaining in vessel)

or put another way

 30 (perfect vacuum) − 16 (gage reading in inches) ÷ 2 =
 7 psi (pressure remaining in vessel)

As you can see, even though the pressure remaining in the vessel is well below that of the surrounding atmosphere, the vessel still contains pressure. Externally it is being subjected to 15 pounds of pressure on every square inch of its surface, while internally the vessel is being pressurized to the tune of 7 pounds on every square inch of its surface. The vacuum scale reads from zero (atmospheric pressure) to 30 (perfect vacuum).

Figure 5-4. Compound Bourdon-Tube Gauge

Absolute Pressure — encompassing the whole range of pressure is the absolute scale. For purposes of mathematical calculation it can be expressed as atmospheric pressure plus gage pressure. Its scale starts at zero (perfect vacuum) and extends forever to infinity.

SPECIFIC HEAT TRANSMISSION

If you were able to decipher the pressure problems, you're one up on most of your peers, but even the headiest of us have difficulty with heat theory. What's all the fuss about? Well, first off: are you aware that heat and temperature are not one and the same thing? Well, I'll be . . ., that's right; heat is thermal energy and temperature is a measure of how hot or cold an object is. You've been reading, haven't you? But did you know that there are two kinds of heat, known by their effects? Figure 5-5 illustrates the relationship between them, using a pound of water at atmospheric pressure in the example.

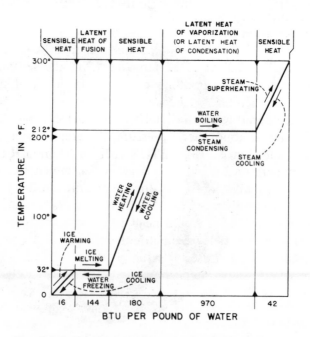

Figure 5-5. Heat Relationships at Atmospheric Pressure

Just as pressure is pressure, heat is heat. The differences we witnessed in the preceding graph deal not with two kinds of heat, but with two of heat's effects. Let's study them to find out why the differences exist.

Sensible Heat — when the flow of heat into or from a substance occurs without benefit of a change of its physical state, we say that sensible heat has been added to or removed from the substance. We can substantiate the transfer through the use of thermometers or other temperature-sensitive devices which indicate corresponding increases or decreases in the temperature of the substance on their scales.

Latent Heat — sometimes referred to as "hidden heat," latent heat is evident only when the physical state of a substance is altered, as when water freezes or ice thaws. Tremendous amounts of heat must be transferred to accomplish changes of state, but your temperature-monitoring equipment will do you little good here, as no rise or fall in temperature takes place during the process. And though the phenomenon is difficult to comprehend, it serves as the basic operating premise for both the steam generating and refrigeration fields.

As you can surmise from the graphic, temperature rises and falls when heat is added to or removed from the sensible heat labeled areas and large amounts of heat addition and removal are required to facilitate a change of state in the water, with no change in temperature in the latent heat labeled areas. Using 1 pound of water at atmospheric pressure, the graph illustrates that it takes the addition of 1354 BTU's (British Thermal Units) to raise the temperature of ice at 0 degrees Fahrenheit to steam at 300 degrees Fahrenheit, 240 of which represent sensible heat and 1114 of which represent latent heat. If the process were reversed, the same amount of heat would need to be removed to cool 300-degree steam to ice at 0 degrees Fahrenheit.

You look a bit confused. Certainly; it makes perfectly good sense to review those two items here.

British Thermal Unit — A B.T.U. is the amount of heat required to raise or lower the temperature of 1 pound of water, 1 degree Fahrenheit.

Specific Heat — the number of B.T.U.'s of heat required to raise or lower 1 pound of substance, 1 degree Fahrenheit. Water, being so abundant, is used as the basis for measurement having a specific heat of 1.0. Cast iron's is 0.1298 and ice has a specific heat value of 0.504; though for purposes of easy calculation it is obvious that a value of 0.5 was used. Can you interpolate the figures and tell what value was used for steam in the graph?

TEMPERATURE SCALES

You say you thought the subject of heat was confusing? When you explained your understanding of the difference between heat and temperature you were on such a roll and sounded so profound, that I didn't have the heart to inform you that you were only half right. True, heat is thermal energy, but temperature isn't exactly a measurement of how hot or cold a substance is. Actually all substances, even the "coldest," have some heat within them. Cold is nothing more than a relative term invented for the masses to help them perceive the thermal energy contained in substances through the human sense of touch. Heat is manifested within a substance by the vibration of its molecules. The faster the molecules vibrate, the more heat a substance contains and, conversely, the slower the molecules vibrate, the less heat it contains.

As we've already learned in the section on heat, a change in the physical state of a substance is a visual indication that latent heat transfer has taken place within a body. Temperature-sensitive devices tell us when sensible heat has been transferred and their temperature scales literally indicate to what degree.

Figure 5-6 represents the scales commonly used throughout the world for measuring temperature.

Certain points on the scales are important to us in power plant engineering; specifically boiling and freezing temperatures. Again, water, under atmospheric pressure conditions, will be used to illustrate these points. Lines have been drawn across the scales where they register the boiling and freezing temperatures of water and absolute zero for comparison. Before you ask . . .

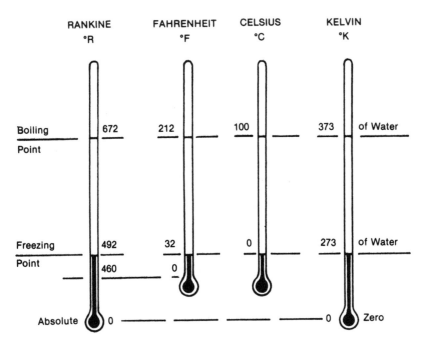

Figure 5-6. Temperature Scales
(Courtesy: Building Owners and Managers Institute)

Absolute Zero — though science has never accomplished the complete removal of all the heat from a substance, it is a generally accepted theory that "absolute zero" is a hypothetical temperature value assigned to a condition characterized by the complete absence of heat in a substance when, it is thought by some, all molecular activity ceases. Absolute zero on the Fahrenheit scale registers at approximately —460 degrees and on the Celcius scale at approximately —273 degrees.

Rankine Scale — an absolute-temperature scale, the unit of measurement of which equals a Fahrenheit degree. It registers the freezing and boiling points of water at 492 and 672 degrees respectively.

Fahrenheit Scale — a thermometric scale on which the boiling and freezing points of water register at 212 and 32 degrees above its zero mark, respectively, under standard atmospheric pressure at sea level.

Celsius Scale — a thermometric scale on which the boiling and freezing points of water register at 100 and 0 degrees respectively, under standard atmospheric pressure at sea level.

Kelvin Scale — an absolute temperature scale, the unit of measurement of which equals a Celsius degree. It registers the freezing and boiling points of water at 273 and 373 degrees respectively.

You're starting to scare me. It seems the more I explain to you, the more you want me to explain to you. Now don't think for one instant that I have less than a full confidence in your potential to handle this job, but sooner or later you're going to have to come up with more answers than questions. I guess it follows . . . if you've got the time, this is the place.

Although we resolve many of our plant problems by reacting intuitively to adverse conditions, applying deductive reasoning to irrational situations or diagnosing them with sophisticated electronic instruments, more often than not, a pencil, scratch pad and a few quiet moments are all you'll need to quell most of the concerns you'll have as an operating engineer. Let's review some of the more common mathematical quandaries which may confront you. I'll provide you with a rule to assist you in solving each problem posed.

TEMPERATURE CONVERSION

Rule
To convert degrees on the Fahrenheit scale to their equivalent Celsius temperatures, use the formula, $C = 5/9 \times (F-32)$

To convert degrees on the Celsius scale to their equivalent Fahrenheit temperatures, use the formula, $F = 9/5 \, C + 32$

Problem
What is the temperature equivalent of —40 degrees Celsius on the Fahrenheit scale?

Solution
9/5 x C (deg. Celsius) + 32 =
(9/5 x −40) + 32 =
(−72) + 32 =

Answer: −40 degrees Fahrenheit

BOILER HORSEPOWER

Rule
One boiler horsepower is equivalent to the evaporation of 34.5 pounds of water per hour at 212 degrees Fahrenheit

Problem
Find the horsepower rating of a vessel which evaporates 430 gallons of water per hour at atmospheric pressure.

(given) 1 gallon of water weighs 8 pounds

Solution
430 (gal) x 8 (lbs/gal) ÷ 34.5 (lbs/hr/bhp) =

Answer: 99.71 or 100 boiler horsepower

DEGREE OF SUPERHEAT

Rule
When a liquid is boiling and generating a vapor it is referred to as a saturated liquid and the vapor is referred to as saturated vapor because each is "saturated" with as much heat as they are capable of containing at the pressure they are under.

The temperature at which a liquid boils is called the saturation temperature and the corresponding pressure is called the saturation pressure. As saturation temperatures rise and fall, saturation pressures rise and fall correspondingly.

At any given pressure, liquid and vapor temperatures are the same and the temperature of one cannot be raised without raising the other as long as they remain in contact.

Superheating is the act of raising a vapor's temperature above the temperature of the liquid from which it was generated by breaking the physical bond between the two and adding additional heat to the vapor at the same pressure. The difference between the saturation temperature of the liquid and the superheated temperature of the vapor is referred to as the Degree of Superheat.

Problem
What kind of superheat is attained when steam is heated from a saturation temperature of 486 degrees Fahrenheit to a superheated temperature of 750 degrees Fahrenheit?

(given) 600 psia constant pressure

Solution
750 (superheated temp) − 486 (saturation temp) =

Answer: 264 degrees of superheat

POWER FACTOR

Rule
In electrical terms, power factor is the ratio of true power to apparent power expressed as a percentage which indicates the efficiency of a circuit.

In direct current (dc) circuits the power factor value is always 1.0 or 100 percent and their power equation is represented as:

$$WATTS = VOLTS \times AMPERES$$

In alternating current (ac) circuits power factor values vary between 0 and 1.0 or zero and 100 percent, depending on load and circuit characteristics and their power equation is represented as:

$$WATTS = VOLTS \times AMPERES \times POWER\ FACTOR$$

Problem
Determine the power factor of a 1200 kva diesel-driven generator that develops 1080 kilo-watts of power at full load.

(given) K or kilo equals 1000

Solution

pf (power factor) = kw (kilowatts) / kva (kilovolt amperes)
                 = (1080 x 1000) / (1200 x 1000)
                 = 1080 / 1200

Answer: 0.9 or 90 percent

TON OF REFRIGERATION EFFECT

Rule

A ton of refrigeration is the equivalent effect of melting one ton (2000 lbs) of ice over a period of 24 hours.

The effect relies on the removal of latent heat to facilitate a physical change since no change in temperature takes place during the fusion process.

Problem

What is the capacity of a refrigeration unit which is capable of extracting 90,000 btu's of heat per hour?

(given) latent heat equals 144 btu's per pound of ice

Solution

a) 1 ton (ref. effect) = 144 (btu's latent heat) x 2000 (lbs. ice)
                       = 288,000 (btu's per 24 hours)
                       = 12,000 (btu's per hour)

b) 90,000 btu's extracted/hr) / 12,000 (btu's/hr) =

Answer: 7.5 tons

POSITIVE DISPLACEMENT

Rule

In reciprocating pumps, a definite quantity of liquid, equivalent to the volume of each cylinder, is displaced with each stroke of the piston contained within it.

Problem

How many gallons per minute can a reciprocating pump discharge if each of its water cylinders has a capacity of 785.4 cubic inches and each piston travels through 24 strokes per minute?

(given) 1 gallon equals 231 cubic inches

Solution

785.4 (cu. in) x 2 (cyl) x 24 (strokes/min) / 231 (cu. in/gal) = 37699.2 / 231 =

Answer: 163.2 or 163 gallons per minute

COEFFICIENT OF PERFORMANCE

Rule

In refrigeration work, the coefficient of expansion (cop) is the ratio of the output, or heat absorbed in a system, to the energy input used to produce the effect.

Problem

What is the COP of a heat pump having an output of 60,000 Btu's per hour derived from an energy input of 3.74 kilowatts?

(given) 1 kw equals 3413 Btu's

Solution

COP = BTU output / equivalent BTU input
    = 30,000 (Btu/hr) / 3740 (watts) x 3.413 (Btu/watts)
    = 30,000 / 12,764.62

Answer: 2.35 (COP)

DEGREE DAYS

Rule

Degree days are values used to denote coldness of the weather over prolonged periods, measured against a standard 65 degrees Fahrenheit over a 24-hour period.

Mean temperatures are derived by adding the high and low temperatures for the day and dividing by 2.

The formula used to determine degree days is 65 degrees minus the mean temperature for each day times the number of days in the period.

Problem

Determine the number of degree days for a 2-day period on which the high and low temperatures were 40 and 60 and 44 and 60 degrees respectively.

Solution

a) 65 (degrees) − 50 (mean temp)

b) 65 (degrees) − 52 (mean temp) =
    a + b = 15 + 13 =

Answer: 28 degree days

Chapter 6

OPERATOR AIDS

It wasn't long ago that we engineers were considered an ignorant lot. I think our problem was image based. It used to be that when people peered down into the boiler room it was hard for them to perceive what was looking back. We were sweaty, soot-covered mounds of flesh armed with little more than coal shovels, rags in our hip pockets and a sincere collective desire to keep our engines running. We didn't know a file cabinet from a foul ball and the only paper work we did was the crossword puzzle we clipped from the local tribune. But that was yesterday; references to us as grease monkeys are remarks from a bygone era. Although it's true our commitment to keep the plant operating has remained our primary objective, we now use those rags for little else than buffing the scuff marks from our shoes and our coal shovels have been replaced with computers.

Today, the operation of a power plant is becoming more of a science than a trade. Equipment is being manufactured to more exacting specifications and to closer tolerances. Many of today's operating controls now contain solid-state electronic boards. Computerized energy management monitors automatically start, modulate and shut down system components on demand. And our job is to make it all work like a Swiss watch. Are you worried? Don't sweat it. Here are some items I've used to stay on top of things over the years. They worked for me; they'll work for you.

STANDARD OPERATING PROCEDURES

Whether they call it the Standard Operating Procedures (SOP) Manual, the Policy and Procedures Guide or the company rule book, every engineering department has an operations bible. The "book" quotes the almighty in the corporate hierarchy, provides counsel to the workers on their conduct within the "hallowed" company walls and contains insights that help its "believers" to accomplish the minor miracles they are charged with performing each day. It has been argued, however, that down in the trenches, it's not an operations bible we need so much as a survival manual. Figure 6-1 displays the Table of Contents of a typical SOP manual. Though all of the information it contains is necessary and relevant to the department's overall responsibilities, only certain of its items are of strict importance to us in our daily work.

In our example, the ORGANIZATIONAL, FINANCIAL and PERSONNEL sections of the SOP manual hold no particular consequence for our day-to-day operations; the OPERATIONAL section covers those items that affect us more than the things affected by us and the EQUIPMENT section spells out the official operating and testing of all the equipment located throughout the physical plant. The PREVENTIVE MAINTENANCE section, although a part of the SOP manual here, is almost always an entity unto itself, covering all aspects of equipment conservation care. The remaining three sections, SAFETY, FIRE and EMERGENCY deal with data you need to keep the plant operating smoothly and safely during normal as well as adverse conditions.

I heartily suggest that if you review no other document in your department, you make it the SOP manual. Much thought, effort and time was put into its construction. It can save you much thought, effort and time it you use it constructively.

A/E RENDERINGS

Looking around, you may not be overly impressed with the architectural design of your building or the layout of its equipment. But no matter; I've yet to meet the operating engineer who is perfectly content in or satisfied with his physical plant. Just be thankful that after it was constructed, the architect and engineers

Figure 6-1. S.O.P. Manual: Table of Contents

```
FIGURE 6-1. S.O.P. MANUAL: TABLE OF CONTENTS

    SECTION                  SUB-SECTION

    ORGANIZATIONAL           - Scope/function
                             - Objectives
                             - Organizational Charts
                             - Management by Objectives Guidelines
                             - Department policy/procedure list
                             - Policies
                               • OPP 2.2  Organizational policies/procedures
                               • MNT 1.0  Maintenance section organization
                               • MNT 1.1  Maintenance section standards

    FINANCIAL                - Policies
                               • OPP 1.44 Capitalization
                               • OPP 1.48 Purchase order
                               • OPP 1.50 Blanket purchase order
                               • OPP 1.46 Equipment acquisition
                               • PO-013   Contracts manual
                             - Vendor list
                             - Department Budgets

    PERSONNEL                - Policies
                               • PO-002   Employee orientation
                               • OPP 1.23 Orientation program
                               • OPP 2.30 Job description
                               • OPP 2.40 Seminar and workshop
                               • OPP 2.7  Seminar attendance
                               • OPP 2.8  Inservice attendance
                               • MNT 1.4  Maintenance dept. training
                             - Job Descriptions
                             - Safety Presentation

    OPERATIONAL              - Policies
                               • MNT 2.0  Duties of maintenance dept.
                               • PO-003   Department work hours
                               • PO-004   Work requisition system
                               • MNT 2.5  Extreme weather preparation
                               • MNT 3.0  Work dress uniform
                             - Procedures
                               • PO-005   Winter groundskeeping
                               • PO-010   Shop/job cleanup
                               • PO-001   Maintenance cart rounds routine
                               • MNT 1.3  Maintenance department security
                               • PO-017   Water treatment program
                             - Schedules
                             - Forms
                             - Drawing Index

                             - 1 -
```

were required to leave some information behind to help you in your effort to maintain it; specifically, a "Statement of Construction," a set of "as built" drawings and a "Specifications Manual." The statement of construction gives you insight into the fire integrity, material construction and compartmentilization of your building's interior structures and the as-built drawings show the corrected routing and location of electromechanical conduits and devices. The specifications manual or closeout file is a ready reference for searching out obscure details of almost any item in the plant.

Figure 6-1 (Continued)

```
SECTION                    SUB-SECTION

EQUIPMENT                  - Policies
                             * PO-006   Emergency generator testing
                                        schedule
                             * MNT 2.4  Freeze protection
                           - Procedures
                             * PO-007   Emergency generator testing
                             * PO-008   Operating instructions -
                                        generator
                             * MNT 2.27 Cleaning of incinerator

PREVENTIVE MAINTENANCE     - Policies
                             * PO-011    Preventive maintenance program
                           - Procedures
                             * MNT 2.25 Humidity control guidelines
                             * PO-011.1 Preventive maintenance rounds
                             * PO-011.2 Mechanical rounds routine
                             * MNT 2.22 Lawn equipment
                             * PO-011.3 Preventive maintenance -
                                        generator
                             * MNT 2.20 Dietary equipment
                             * MNT 2.21 Vehicles

SAFETY                     - Policies
                             * OPP 1.17 General safety rules
                             * OPP 1.24 Safety program
                             * OPP 2.21 Risk management/safety committee
                             * PO-014   Department safety
                           - General Procedures
                             * OPP 2.37 Safety inspection
                             * OPP 2.51 Hot water temperatures
                           - Electrical Procedures
                             * OPP 1.26 Electrical safety program
                             * OPP 2.60 Use of personal electrical
                                        appliances
                             * OPP 2.47 Extension cords/adapters
                             * PO-016   Department electrical safety

FIRE                       - Policies
                             * MNT 2.6  Maintenance of automatic
                                        sprinkler system
                             * OPP 1.25 Fire drill
                             * MNT 2.7  Fire extinguisher
                                        inspection/testing
                             * OPP 1.4  Smoking

                              - 2 -
```

MANUFACTURERS' PUBLICATIONS

Every piece of equipment made in America comes replete with reams of supporting documents and volumes of instructional materials. Infrequently, the paperwork accompanying the unit is given to the operating engineer. More often than not it is shuffled upstairs with the packing slip and relegated to a file drawer, never again to see the light of day. Worse than that, sometimes it's thrown out with the packing crate. There is good reason for being

Figure 6-1 (Continued)

```
        SECTION                    SUB-SECTION

                                ♦ MNT 2.19 Fire warning system
                                ♦ OPP 2.35 Use/storage of flammable
                                           liquids
                                - Procedures
                                ♦ OPP 1.35 Fire procedure

        EMERGENCY               - Policies
                                ♦ OPP 1.34 Provision for essential
                                           services
                                - Procedures
                                ♦ OPP 2.52 Evacuation of personnel
                                ♦ PO-015   Equipment contingency plan
                                ♦ OPP 1.38 Bomb threat
                                ♦ PO-015.1 Elevator evacuation
                                - Telephone numbers



                                   - 3 -
```

deliberate in the processing, distribution and retention of manufacturers' literature. Since you're new at this, I've written down some guidelines you might wish to consider; to wit:

- registration cards should be filled out to alert manufacturers to dates when units are placed into service thereby validating their guarantees.

- you should note the warranty dates, conditions and exclusions to keep from voiding them as the result of untimely or unwarranted repairs.

- electrical schematics and pictorial diagrams should be contained in transparent plastic pockets that are readily accessible for trouble shooting, easily cleaned and stored for re-use.

- service manuals should be kept in a clean, dry place and completely reviewed before attempting to employ their tactics undertaking repairs.

- maintenance recommendations should be incorporated into your company's preventive maintenance program and their procedures strictly adhered to.

- parts manuals should be retained in the engineering office and referred to for ordering consumable stock items and replacements.

- design and performance specifications indicating recommended tolerances, limits and capacities should be thoroughly understood before attempting to modify or repair any device or system component.

WRITTEN PROCEDURES

No engineer can keep every bit of information pertaining to his job in his head. There is a staggering amount of data to consider in operating even the most modest of physical plants and a prescribed way of preforming every task. There are procedures for starting, operating, shutting down, cleaning, inspecting, installing, testing, dismantling, lubricating, and repairing the equipment, devices and system components in all physical plants. The more enlightened of us write them down.

Why? Written procedures detailing the steps to be taken in starting, operating and stopping equipment are often complex due to the sequential order in which the steps must be taken. This complexity is compounded when long periods of time elapse between implementation of procedures. The operating instructions in Figure 6-2 and the testing procedures in Figure 6-3 are good illustrations of this.

Figure 6-2. Generator Operating Instructions

Policy Number PO-088

Effective Date 12/8/84

I.   Originator: Director; Plant Operations

II.   Title: Operating Instructions - Generator

III. Scope: Plant Operations Department

IV.   Purpose: To ensure proper operation of the emergency power

                  diesel generator

V.   Text:

PRE-START CHECKLIST

1. Place the control panel selector switch which is labeled "Manual-
   Off-Auto" in the "Off" position.

2. Check generator and electrical rooms for any obvious discrepancies.

3. Check that engine and radiator are free of debris, foreign objects,
   and loose or broken parts.

4. Check fuel level in day tank..

5. Check governor oil level.

6. Check engine oil level with the dipstick.

7. Check coolant level; it should be above the baffle plate, one-half
   inch below the fill pipe.

8. Check fan and alternator belts for wear and tension; deflection
   should be 9/16 to 13/16 inches @ 25 lbs. force.

Figure 6-2 (Continued)

9. Check the oil heater to insure it is operating.

10. Check to insure that the engine, generator, radiator and intake
    and exhaust louvers are free of foreign objects and debris.

11. Examine the motor operated louvers to insure that their linkage
    is in proper adjustment so as to keep the louvers tightly closed
    when the generator set is at standstill.

12. Make certain that all fuel oil valves are in the "open" position.

13. Check the battery water level and measure the specific gravity
    of each cell. Inspect the battery terminals for signs of corrosion
    and tightness of connectors. Clean the top of the battery.

14. Correct any discrepancies before starting engine.

15. Place the control panel selector switch which is labeled "Manual-
    Off-Auto" in the "Auto" position.

16. Make sure the generator breaker is in the "On" position.

                            POWER UP SEQUENCE

1. Pull the main electrical switch into the "Off" position and allow
    the generator time to accept the load.

2. Pull the EC main switch into the "Off" position and allow the
    generator time to accept the load.

3. Pull the EQE main switch into the "Off" position and allow the
    generator time to accept the load.

    NOTE: Load acceptance by the generator should be checked at
            the generator control panel after each main switch is
            pulled. Loading of the generator is accomplished when

Figure 6-2 (Continued)

an increase in amperes is manifested on the gauge.
If transfer does not take place, through the transfer
switch, for each main circuit, the run should be aborted
and the Director of Plant Operations should be contacted
immediately.

OPERATING CHECKLIST

1. Stay with the generator while it is in the operating mode, being
   cognizant of any unusual noises or abnormal vibration of the unit
   which would cause you to abort the run.
2. Using the operating log, record all required readings as you observe
   the units operation.

POWER DOWN SEQUENCE

Return the four main switches to the "On" position in the reverse order
that they were deenergized and observe transfer switches for proper
operation.

POST TEST PROCEDURE

1. Stay with the generator during its idle down period until it shuts
   down completely.
2. Check to insure that control panel switch labeled "Manual-Off-Auto"
   is in the "Auto" position.
3. Check the level in the fuel tank and refill if necessary.
4. Inspect for proper closing of the intake and exhaust louvers.

Figure 6-2 (Continued)

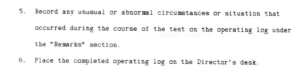

    5.   Record any unusual or abnormal circumstances or situation that
         occurred during the course of the test on the operating log under
         the "Remarks" section.

    6.   Place the completed operating log on the Director's desk.

                                                ————————————————————

                                              Originator   Title   Date

                                              ————————————————————

                                              Approval   Title   Date

Figure 6-3. Water Testing Prodecure

Policy Number   <u>PO-011</u>

Effective Date   <u>1/4/76</u>

I.   Originator: Director; Plant Operations

II.   Title: Boiler Water Test Procedures

III.   Scope: Plant Operations Department

IV.   Purpose: To maintain boiler water values within

                   recommended parameters

V.   Text:

Requirement

Solids concentrations shall be controlled such that the
boiler water conductivity is maintained between 4000 and
5000 mmho

Test Procedure

* Pour cooled boiler water sample into bowl
* Bring meter in range to x 1000
* Multiply the reading by 1000
* Record findings on log sheet
* Increase/decrease blowdown as needed to bring
  reading into line

Figure 6-3 (Continued)

Requirement

Sodium sulfite shall be used as an oxygen scavenger

maintained at a level between 40 - 75 ppm

Test Procedure

* Pour 100cc of cooled boiler water into a clean bowl
* Add 2cc of stable starch, 5cc of dilute hydrochloric

    acid and 5cc of potassium iodide
* Slowly add iodate solution from burette while stirring
* Stop when the water turns blue
* Multiply the burette cc reading by 10 and record on the

    log sheet
* Add sulfite as needed to bring value into line

Requirement

Phosphates shall be maintained within the range of 35-70 ppm

in order to control the formation of scale on the boilers

heating surfaces

Test Procedure

* Pour a filtered sample of cool boiler water up to the first

    mark on the graduated cylinder
* Add enough phosphate reagent #2 to bring the level to the

    second mark
* Cover the mouth of the cylinder and shake well

Figure 6-3 (Continued)

* Add enough reagent #3 to bring level up to the third mark
  and repeat mixing procedure
* Let stand 5 minutes and compare with the comparator
* Record reading on the log sheet
* Increase/decrease phosphate levels until value is brought
  into line

Requirement

The alkalinity of the water shall be maintained at a value
between 8.0 and 9.5 pH. to control corrosion of the boilers
watersides

Test Procedure

* Pour 100cc of cooled boiler water sample into a bowl
* Add 10cc of neutral barium chloride
* Add phenolphtholein solution until water turns pink
* Add hydrochloric acid from burette until pink color
  disappears
* Record burrette reading on log sheet
* Increase/decrease caustic soda amounts until readings are
  brought into line

CODE BOOKS

Someone once said that opinions were like noses; everyone has one of their own. That thought probably wasn't coined with the Stationary Engineer in mind, though we are an opinionated lot, since much of what we do is regulated by others and not subject to our individual whim or decision. Granted we don't like other people sticking their noses into our business, but I think we all agree that some regulation in our field is necessary if not welcome, to aid in its safety, standardization and technical progression. In order to properly comply with the dictates of the regulatory bodies under whose jurisdiction we fall, the operating engineer should have access to at least the following:

- National Electrical Code
- Life Safety Code
- National Fire Codes
- Building Codes
- Selected A.S.M.E. Code sections
- Local ordinances affecting the operation

THE ENGINEERS' LIBRARY

We all like to be considered the "last word" when it comes to the subject of power plant operations and quite honestly, many in our ranks can be considered truly expert in that regard but there's not one of us that hasn't been found wanting for the right answer at one time or another in our careers. Aside from a bruised ego, lacking the appropriate answer during the course of general discussion is of no consequence. It's when you're alone on the job and the place is falling down around you that the most harm can be done. If you don't possess the answers yourself, you'd better have them close at hand. One way of assuring success in the face of adversity is by referencing technical publications. Every plant should have an engineers' library and every engineer should be encouraged to use it. The collection should include but not be limited to:

- handbooks on engineering fundamentals
- an automotive encyclopedia

- technical dictionaries
- manuals on building construction and care
- handbooks on the subjects of air conditioning and refrigeration, boilers and pressures vessels, engines, motors, pumps, air handling systems and fan control, . . . etc.
- trade books covering the fields of carpentry, plumbing, welding, masonry, steam fitting and plant maintenance . . . etc.
- texts on the topics of electricity, pneumatics, hydraulics . . . etc.
- a copy of this book

VENDOR ROSTERS

At times you'll feel deserted when you're down here by yourself for hours on end, but take comfort in the fact that you're not alone in your isolation. Many of our company's vendors are accessible on a 24-hour-a-day basis and the people manning their telephones after hours probably feel as foresaken as you. Of course that reason itself doesn't justify making contact, but it's reassuring to know that someone's out there if you need them. Your office should maintain a file on all vendors with whom they presently do or have ever done business, which should contain this information:

- the vendor name and headquarter address
- a list of available services
- local addresses and telephone numbers
- names of vendor contacts and their titles
- local, toll-free and emergency telephone numbers
- copies of existing contracts
- a product line and list of distributors
- hours and days of operations
- problem service histories involving your plant

EQUIPMENT HISTORY FILES

Before making a decision as to how you'll tend any piece of equipment or device in your plant, give some thought to its age, history of operation and incidence of repair. A well documented file should be kept on every machine and system component in your charge, for one very simple reason; namely, for 16 hours a day they are in someone else's charge. And 2 days out of the week you don't see them at all. Besides, it's hard enough remembering your anniversary, much less recalling when and what you did to a fuel oil pump last June or September. What should be recorded there? At least:

- the unit building and system location
- model and serial numbers
- manufacturer's name, address and telephone number
- equipment specifications
- warranty information
- date and from whom purchased
- date installed and put into service
- location of parts and service manuals
- preventive maintenance & testing requirements
- nameplate data
- numbers assigned by regulatory bodies
- cross reference to building drawings
- inspection dates and findings
- chronological narrative of repairs
- inventory of parts in stock

REPORTING FORMS

I know, they're a pain in the . . . , no it's not that they don't trust you to . . . , sure it's difficult to juggle clipboards and . . . , wait a minute. Taking readings has been a sore spot with operating engineers for years but there's a perfectly good reason for doing it.

You know if your equipment is functioning properly when you give it the once-over during your mechanical rounds, but what about the other shift engineers, machinery repair people and your boss, as far as that goes? Equipment readings taken over long periods of time are used to identify operating trends, track temperature and pressure fluctuations, flag nuisance tripping episodes, indicate equipment status, record hours of operation and alert appropriate personnel to problems in the physical plant that might otherwise go unnoticed or unreported. Figures 6-4 through 6-6 are examples of the types of forms typically used for recording equipment readings in the power plant. Forms are also used to report findings such as discrepancies in the power plant as illustrated in Figure 6-7 and for documenting water test results.

Figure 6-4. Operating Log

PLANT OPERATIONS   10/84

EMERGENCY GENERATOR OPERATING LOG

Date	RUN TIME		TANK LEVEL		LOADING			TEST ABORT		REASON
	Start	End	Day	Fuel	Sim	Outage	Full	Yes	No	

PRE-START

OIL LEVEL		BATTERY		FUEL TANK	COOLANT LEVEL		FILTERS		OPERATION	
OK	?	SG.	Lev.	Gallons	OK	?	Air	Oil	Hours	Last Date

OPERATING

VOLTS	CYCLES	AMPERES			WATER	OIL	FUEL	INDICATORS		TIME LAG			
		Leg. 1	Leg 2	Leg 3	Temp	PSI	PSI	Oil	Air	TS-1	TS-2	TS-3	TS-4

POST-TEST

OIL LEVEL		BATTERY		FUEL TANK	COOLANT LEVEL		FILTERS		OPERATION	
OK	?	SG.	Lev.	Gallons	OK	?	Air	Oil	Hours	Diff

REMARKS

_____

_____

_____

OPERATOR SIGNATURE _____

Figure 6-5. Operating Log
(Carrier Corporation — Division of United Technologies Corporation)

Carrier Air Conditioning

Carrier Parkway
Syracuse, New York 13221

Division of
Carrier Corporation

Carrier **Refrigeration Log**

PLANT _____

DATE _____

CARRIER 19D HERMETIC CENTRIFUGAL REFRIGERATION MACHINE

MACHINE SERIAL NO. _____   MACHINE SIZE _____

	COOLER				CONDENSER					COMPRESSOR							PURGE			
	GPM					GPM						OIL								
Time	Vacuum	Refrigerant Temperature	Refrigerant Level	Water Temperature IN	Water Temperature OUT	Vacuum or Pressure	Condensing Temperature	Water Temperature IN	Water Temperature OUT	Bearing and Transmission Temperature	Level	Temperature	Pressure	Reservoir Pressure	Motor Amps or Vane Position	Refrigerant Level	Condenser Pressure	Frequency of Pump Operation	Refrigerant Pump On or Off	Operator's Initials
1	2	3	4	5	6	7	8	9	10	11	12	13	14	15	16	17	18	19	20	21

REMARKS: Indicate shut downs on safety controls, repairs made, oil or refrigerant added or removed, and water drained from purge. Include amounts.

Figure 6-6. Operating Log

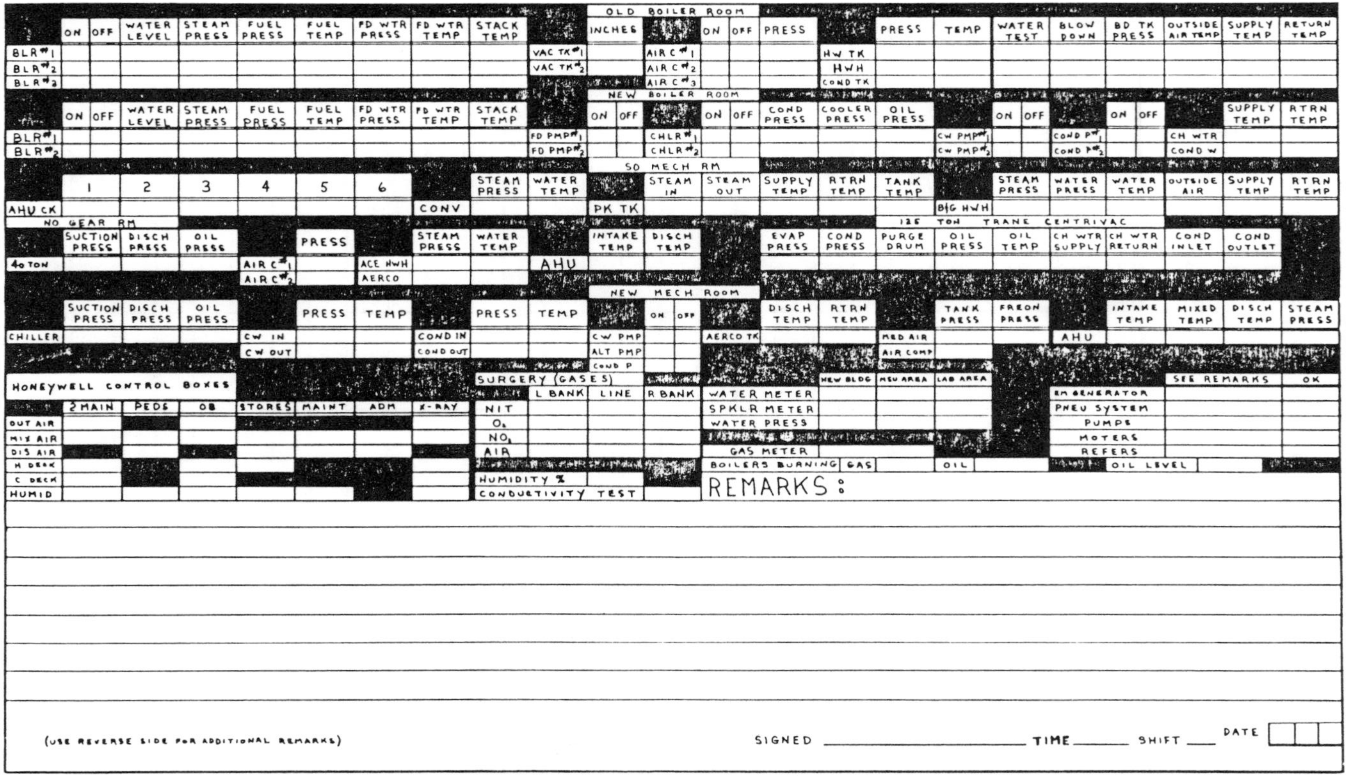

OLD BOILER ROOM

ON | OFF | WATER LEVEL | STEAM PRESS | FUEL PRESS | FUEL TEMP | FD WTR PRESS | FD WTR TEMP | STACK TEMP | INCHES | ON | OFF | PRESS | PRESS | TEMP | WATER TEST | BLOW DOWN | BD TK PRESS | OUTSIDE AIR TEMP | SUPPLY TEMP | RETURN TEMP

BLR#1 | BLR#2 | BLR#3 | VAC TK#1 | VAC TK#2 | AIR C#1 | AIR C#2 | AIR C#3 | HW TK | HWH | COND TK

NEW BOILER ROOM

ON | OFF | WATER LEVEL | STEAM PRESS | FUEL PRESS | FUEL TEMP | FD WTR PRESS | FD WTR TEMP | STACK TEMP | ON | OFF | ON | OFF | COND PRESS | COOLER PRESS | OIL PRESS | ON | OFF | ON | OFF | SUPPLY TEMP | RTRN TEMP

BLR#1 | BLR#2 | FD PMP#1 | FD PMP#2 | CHLR#1 | CHLR#2 | CW PMP#1 | CW PMP#2 | COND P#1 | COND P#2 | CH WTR | COND W

SO MECH RM

1 | 2 | 3 | 4 | 5 | 6 | STEAM PRESS | WATER TEMP | STEAM IN | STEAM OUT | SUPPLY TEMP | RTRN TEMP | TANK TEMP | STEAM PRESS | WATER PRESS | WATER TEMP | OUTSIDE AIR | SUPPLY TEMP | RTRN TEMP

AHU CK | CONV | PK TK | BIG HWH

NO GEAR RM | 125 TON TRANE CENTRIVAC

SUCTION PRESS | DISCH PRESS | OIL PRESS | PRESS | STEAM PRESS | WATER TEMP | INTAKE TEMP | DISCH TEMP | EVAP PRESS | COND PRESS | PURGE DRUM | OIL PRESS | OIL TEMP | CH WTR SUPPLY | CH WTR RETURN | COND INLET | COND OUTLET

40 TON | AIR C#1 | AIR C#2 | ACE HWH | AERCO | AHU

NEW MECH ROOM

SUCTION PRESS | DISCH PRESS | OIL PRESS | PRESS | TEMP | PRESS | TEMP | ON | OFF | DISCH TEMP | RTRN TEMP | TANK PRESS | FREON PRESS | INTAKE TEMP | MIXED TEMP | DISCH TEMP | STEAM PRESS

CHILLER | CW IN | CW OUT | COND IN | COND OUT | CW PMP | ALT PMP | COND P | AERCO TK | MED AIR | AIR COMP | AHU

HONEYWELL CONTROL BOXES | SURGERY (GASES) | NEW BLDG | MSU AREA | LAB AREA | SEE REMARKS | OK

2MAIN | PEDS | OB | STORES | MAINT | ADM | X-RAY | L BANK | LINE | R BANK

OUT AIR | MIX AIR | DIS AIR | H DECK | C DECK | HUMID

NIT | O_2 | NO_2 | AIR

HUMIDITY % | CONDUCTIVITY TEST

WATER METER | SPKLR METER | WATER PRESS | GAS METER | BOILERS BURNING | GAS | OIL

EM GENERATOR | PNEU SYSTEM | PUMPS | MOTERS | REFERS | OIL LEVEL

REMARKS:

(USE REVERSE SIDE FOR ADDITIONAL REMARKS)

SIGNED ______ TIME ______ SHIFT ______ DATE

Figure 6-7. Power Plant Inspection Form

PHYSICAL PLANT EVALUATION

POWER PLANT			BUILDING	AREA

ITEM	ACCEPTABLE		REMARKS
	YES	NO	
FUEL LEAKS			
OIL LEAKS			
WATER LEAKS			
STEAM LEAKS			
AIR LEAKS			
SUMP PUMP OPERATION			
FLOOR DRAINS			
DOORS, STAIRS, LADDERS			
WALKWAYS, GRATINGS			
PIPE HANGERS			
EXPANSION JOINTS			
PIPING INSULATION			
EQUIPMENT INSULATION			
AIR LOUVRES, FILTERS			
ROOM LIGHTING			
INDICATOR LIGHTS			
CHEMICAL FEED PUMPS			
WATER PRESSURE			
STACK TEMPERATURES			
UNUSUAL NOISES			
EXCESSIVE VIBRATION			
STAND BY FUEL			
CHEMICAL PAR LEVELS			
PARTS INVENTORY			
ANCILLARY ROOMS			
SURVEYOR			DATE

THE ENGINEER'S LOG

Depending on his religious convictions, probably no written work in recorded history carries more relevance or importance for the Stationary Engineer than his log book. The engineer's log is preferably a hard-bound, sequentially numbered, lined journal into which the operator makes entries documenting the events of his shift. When completed, the log can be used as a legal document for purposes of litigation in a court, as an historical chronology of the plant's operations or, as in my case, as a cherished remembrance of a life's work.

Chapter 7

TOOLS OF THE TRADE

The items in the last chapter can be construed as tools of a sort, but given all the paper in Washington, D.C., you can't diagnose and repair your equipment ills with mere forms, procedural check-lists and written routines; for that you need hardware. Since you've already demonstrated your knowledge and skillful use of the various hand tools we use in the power plant and I assume you're familiar with safety gear, I've decided to forego discussion on those items and break to the quick. Over on the workbench, I've assembled some devices commonly used by operating engineers which I refer to as "the thinking man's tools of the modern electronic age." As you haven't yet worked with most of them and you'll soon be on your own, I thought it might be a good idea for me to school you in their uses. Here's a brief summary of what each of them is used for:

Combustion Analyzer

Measures oxygen content, temperature and carbon-monoxide levels of combustion exhaust gases from which carbon-dioxide levels are interpolated enabling adjustments to be made to increase fuel-burning efficiency.

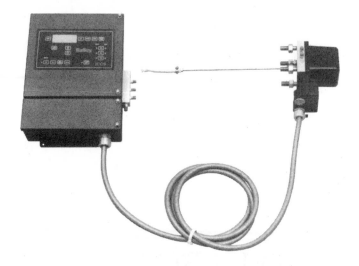

Figure 7-1. Combustion Analyzer
(Courtesy Bailey Controls Company)

Combustible Gas Transmitter

Converts explosive gas concentrations into electrical output for metering, recording or controlling levels.

Figure 7-2. Combustible Gas Transmitter
"Mitchell Instrument, San Marcos, CA"

Dead Weight Tester

*Tests and calibrates
pressure gauges, trans-
mitters, transducers,
recorders, controllers
and receivers in pres-
sure vacuum ranges.*

Figure 7-3. Dead Weight Tester
"Mitchell Instrument, San Marcos, CA"

Dial Thermometer

*Measures temperatures in
the range to which it is
calibrated and is manufactured from special materials
for ultra-high and low
applications.*

Figure 7-4. Dial Thermometer
*"Mitchell Instrument,
San Marcos, CA"*

Digital Thermo-clock

Displays temperature and time in large liquid crystal display numbers and is portable or can be mounted by or on equipment.

Figure 7-5. Digital Thermo-clock
"Mitchell Instrument, San Marcos, CA"

Digital Watt/Watt-Hour Meter

Measures the amount of power required by a complex load, automatically over any period of time and accumulated power consumption in kilowatt hours.

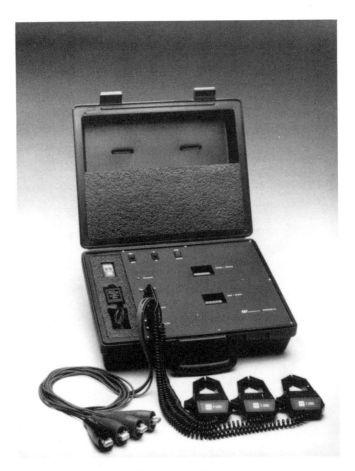

Figure 7-6. Digital Watt/Watt Hour Meter
"Mitchell Instrument, San Marcos, CA"

Electronic Refrigerant Charging Meter

Automatically dispenses refrigerant into systems directly from the suppliers' tanks.

Figure 7-7. Electronic Refrigerant Charging Meter
"Mitchell Instrument, San Marcos, CA"

Fiberscope

Enables visual inspection of otherwise inaccessable areas such as the interiors of pipes, heat exchangers, boilers, machinery components . . . etc.

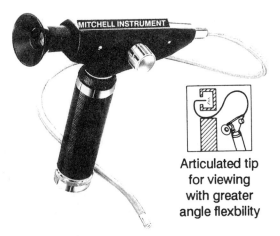

Articulated tip
for viewing
with greater
angle flexbility

Figure 7-8. Fiberscope
"Mitchell Instrument, San Marcos, CA"

Gaussmeter

Measures the strength of magnetic fields.

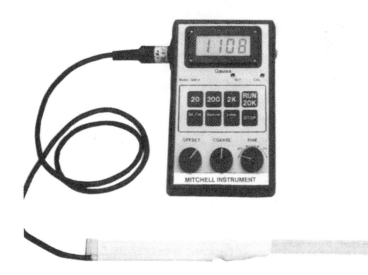

Figure 7-9. Gaussmeter
"Mitchell Instrument, San Marcos, CA"

Halogen Leak Detector

Detects leaks in refrigeration systems, halon fire extinguishing systems and lines carrying mixed halogen-based substances such as ethyleneoxide.

Figure 7-10. Halogen Leak Detector
"Mitchell Instrument, San Marcos, CA"

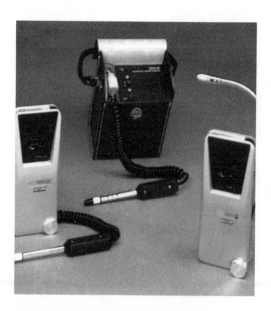

Hermetic Analyzer

Checks motors for starting torque, acid in refrigerants, insulation breakdown, run windings and open relays.

Figure 7-11. Hermetic Analyzer
"Mitchell Instrument, San Marcos, CA"

Hygrometer

Measures the percent of moisture in the air at a given temperature

Figure 7-12. Hygrometer

Infrared Thermometer

Provides instant temperature readouts of warm surfaces taken from a distance such as from steam traps, stacks, boiler furnaces, electrical panels . . . etc.

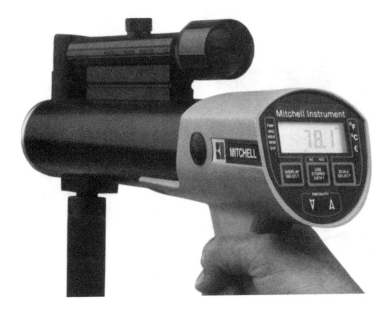

Figure 7-13. Infrared Thermometer (Heat Gun)
"Mitchell Instrument, San Marcos, CA"

Megohmeter
Indicates the insulating integrity of electrical wire coverings

Figure 7-14. Megohmeter
"Mitchell Instrument, San Marcos, CA"

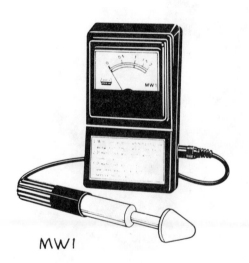

Microwave Radiation Leakage Detector
Detects and measures the amount of microwave energy leakage radiated by ovens, heaters, dryers and other industrial equipment generating high power at microwave frequencies.

Figure 7-15. Microwave Radiation Leak Detector
"Products by Universal Enterprises Inc. and Kane-May"

Multimeters

Measures AC and DC voltage, current and resistance.

Figure 7-16. Multimeters
"Mitchell Instrument, San Marcos, CA"

pH Meters

Measures the acidity or alkalinity levels of boiler and cooling tower waters.

Figure 7-17a. pH Transmitter
"Mitchell Instrument, San Marcos, CA"

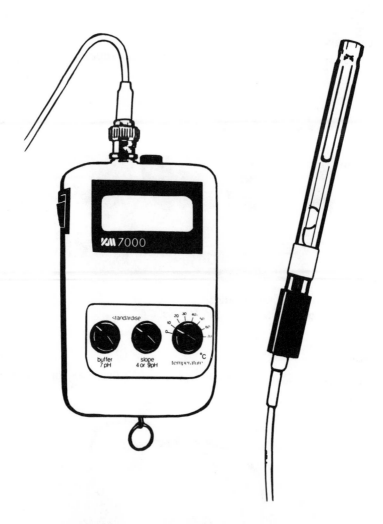

Figure 7-17b. pH Meter
"Products by Universal Enterprises Inc. and Kane-May"

Figure 7-18. Pocket Light Meter

Pocket Light Meter

Measures light volumes in foot-candles and is used for estimating brightness and reflectivity of objects.

Portable Temperature Recorders

Provide continuous permanent records of room temperature variations.

Figure 7-19. Portable Temperature Indicator
"Mitchell Instrument, San Marcos, CA"

Power Factor Clamp-On
Reads power factor values directly from electrical circuit wiring and can be used to determine motor efficiency and loading.

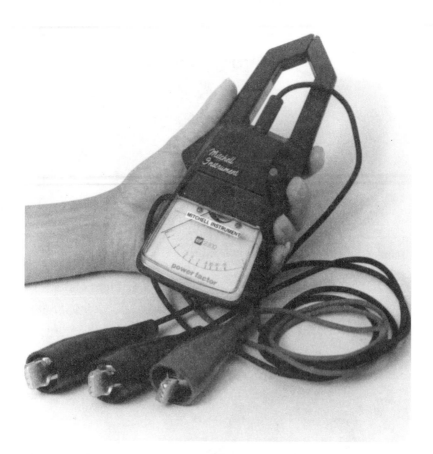

Figure 7-20. Power Factor ClampOn
"Mitchell Instrument, San Marcos, CA"

Pressure Recorder
Measures and records water, air and gas pressures providing a permanent record of variations over time.

Figure 7-21. Pressure Recorder
"Mitchell Instrument, San Marcos, CA"

Refrigeration Charging Gauges
*Reads high and low side refrigerant pressures and is used to charge
systems with new refrigerant.*

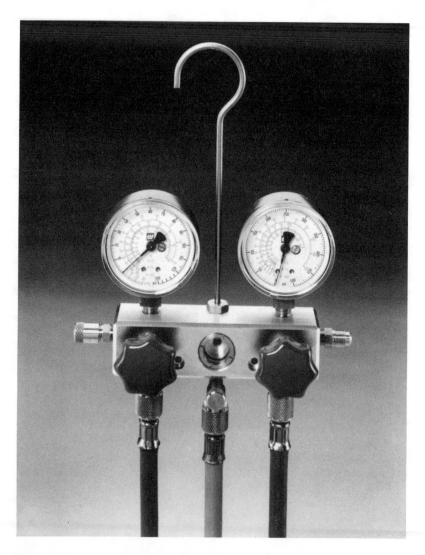

Figure 7-22. Refrigeration Charging Gauges
"Mitchell Instrument, San Marcos, CA"

Remote Reading Temperature Recorder
Records temperatures unmonitored for 24-hour periods in remote areas and can be modified to alarm or control when pre-set values are reached.

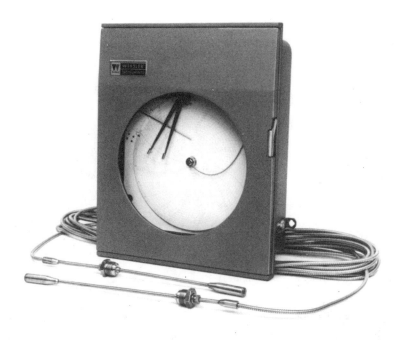

Figure 7-23. Remote Reading Temperature Recorder

Shaft Aligner
Indicates angular and offset misalignments of connecting shafts.

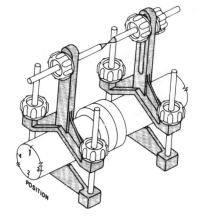

Figure 7-24. Shaft Aligner

Sling Psychrometer

Measures comfort conditions in air conditioned environments and is used to determine the humidity of air associated with critical manufacturing processes in industrial atmospheres.

Figure 7-25. Sling Psychrometer

Smoke Emitters

Generates smoke for leak testing and balancing air and exhaust systems, smoke stacks, pipelines, ductwork and pressure vessels.

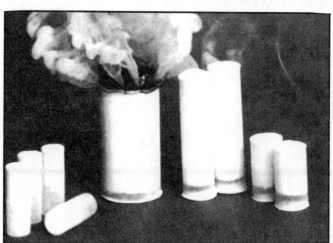

Figure 7-26.
Smoke
Emitters

Stroboscope/Tachometer

Measures speed of motors, gears, shafts, fans, pulleys . . . etc., by rapidly flashing light onto a marked moving surface and matching their frequencies.

Figure 7-27. Stroboscope Tachometer
"Meylan Corporation, New York, New York"

Temperature/Humidity Recorder

Measures and records temperature and humidity values for a 24-hour period.

Figure 7-28. Temperature/Humidity Recorder
"Mitchell Instrument, San Marcos, CA"

Thermocouple Probe

Measures a wide range of temperatures from the minus hundreds to the plus thousands on the Fahrenheit scale and can be purchased with an analog or a digital readout.

Figure 7-29. Thermocouple Probe
"Mitchell Instrument, San Marcos, CA"

Uninterrupted Power Supply (UPS)

Supplies immediate back-up power to circuits for limited periods of time during outages of normal power.

Figure 7-30. Uninterrupted Power Supply (UPS)
"Mitchell Instrument, San Marcos, CA"

Vane Digital Anamometer
Measures air velocity and temperatures in air ducts and from fans, blowers, exhaust hoods and air conditioners.

Figure 7-31. Vane Digital Anemometer
"Mitchell Instrument, San Marcos, CA"

Vibration Analyzer

Measures equipment vibration in displacement, velocity and acceleration.

Figure 7-32a. Vibration Analyzer
"Mitchell Instrument, San Marcos, CA"

Figure 7-32b. Vibration Activated Time Indicator
"Meyland Corporation, New York, New York"

Chapter 8

SUGGESTED MAINTENANCE FREQUENCIES

So you know how to use all the tools and test instruments we own; . . . I'm impressed . . . and you can start up, operate and shut down any piece of equipment in the place; . . . marvelous! . . . and when you take equipment readings you log your findings neatly and accurately; . . . that's ducky. Do you know how and when to pull maintenance on it all? Uh-huh . . . I was afraid you'd say that. I know, I know and I'll even help you set it up, but before you can start a PM program you have to know when to do what to your equipment and you may not know as much about that as you might think. Let's take a tour of the plant and I'll give you an idea of what I'm talking about.

SCHEDULING PROBLEMS

Most systems and equipment included in PM programs adhere to unwavering maintenance schedules based on days, weeks, months, years . . . etc. These types of programs enable you to look into the future and determine what consumable materials and man-hours will be needed to perform preventive procedures on any number of devices during a given period of time. But it's an ideal system and doesn't work well in the real world, due in large part to the influences of my three favorite people; Mr. Murphy, Mother Nature and Father Time.

We've all heard of Mr. Murphy's infamous laws, the gist of which educate us to the fact that if anything can go wrong, it will and at the worst possible time. Applied to our PM program, that equates to getting half way through a preventive maintenance routine and finding that the unit you're working on is in need of repair; a prospect that can completely disrupt your maintenance plans. If Murphy throws schedules out of kilter then Father Time makes it all but unnecessary to have them. And whereas, a well executed care plan can extend the useful life of your equipment, Father Time will eventually assure its removal from your schedule entirely. Mother Nature on the other hand exacts a different kind of toll. The unpredictable old girl butts in when and where she isn't wanted. She alone is responsible for the oddity of seasonal maintenance. Her constant interference and changing moods cause us to regulate our schedules in accordance with the climate instead of by the clock.

SEASONAL SLOWDOWNS

With all the modern technological advances man has made this century, he must still alter his schedules to accommodate inclement weather, abnormal climatic variations and "Acts of God" such as hurricanes, tornadoes and floods. Here we are solely concerned with the effects of weather and climate. I only mentioned the other to show that, as mean as she is, even Mother Nature has to answer to a higher authority.

Changes in the seasons correspond with changes in outdoor temperature, humidity, precipitation patterns and hours of available daylight, causing us to either lock ourselves up in stuffy buildings or flee to the outside for fresh air, depending on the time of year. Of particular concern to the preventive maintenance specialist, these changes affect the operating hours of equipment, interfere with preventive maintenance schedules and cause additional work to be performed on power plant systems and components that might not normally be done, save for the seasonal nature of its operation.

There are devices in your domain, the care of which can be affected by the whimsical ways of Mother Nature. For example, you may be thwarted by snow, wind, rain or ice when attempting to service your building's exhaust fans, roof-mounted air conditioning units or other equipment located outside of your building. The electrical gear in your basement may need to dry out or be baked out after a flood during which time the sump pumps run continuously. Maintenance of light standards and antennae may have to be postponed as the result of high winds, precipitation or lightning in the area.

Two major areas in which seasonal maintenance is conducted are comfort heating and comfort cooling in buildings. The hearts of these systems are their boilers and air conditioning chillers. Review these seasonal generic maintenance requirements for these devices and use them as a guide when you put your PM program together.

Spring Boiler Maintenance
Fall Chiller Maintenance

- Check system piping for leaks
- Test all operating controls
- Shut down and drain units
- Remove and inspect operating controls
- Thoroughly clean unit interiors
- Inspect heat transfer surfaces
- Lubricate all moving parts
- Repack auxiliary valves and pumps
- Lay up per manufacturer's guidelines

Spring Chiller Maintenance
Fall Boiler Maintenance

- Renew gaskets and observation glasses
- Install new filters as required
- Tighten loose connections and fasteners
- Calibrate gages and operating controls
- Replenish fluids per manufacturer's specifications
- Inspect system piping for leaks
- Lubricate all moving parts

- Check operation of auxiliary equipment
- Test safety interlocks

How's that? You're a quick study. Sure I threw you a curve-ball; the further we get into your training, the harder it is for me to teach you anything. Why wouldn't I be able to use the same generic checklist for a chiller as I do for a boiler? Essentially they're the same machine. Both are made of metal and insulated, have moving parts, operating controls and safety devices and both require a myriad of auxiliaries with which to operate. But more significantly, both are heat exchangers. Their difference lies in the fact that one provides heat and the other removes it. I wonder if this correlation was the spark that flamed the idea of the heat pump? Don't you just hate it when you get flashes of brilliance just after someone's already patented the product?

DAILY DUTIES

The first bastion of defense against premature failure of your equipment is being cognizant of its existence. Just as we tend to take our loved ones for granted, due largely to their constant presence at home, at work we come to expect dependable functioning of our machines. As we all eventually learn, some later than others, when we disregard the needs of those around us, the level of their responsiveness to us diminishes. The same holds true for our equipment. And just as a little tender loving care restores the attentiveness of our family members or acquaintances, so it restores the operating integrity of the devices in our charge. Here's a checklist for providing daily TLC to yours. You might want to refer to them when setting up your program.

Daily

- Properly position valves and switches
- Investigate unusual noises and make repairs
- Adjust system pressures and temperatures
- Replenish fluid levels and repair leaks when found
- Move operating controls through their range of motion

- Clean all visual indicating devices
- Measure bearing temperatures and check shaft alignment
- Eliminate problems of excessive vibration
- Lubricate rotating and sliding parts
- Test plant waters and blow down or bleed off units as required
- Drain off condensate from air tanks and lines
- Maintain proper tolerances and clearances
- Replace burned out indicator bulbs and blown fuses
- Tighten all loose connections

WEEKLY WORK

You say your equipment seems to be fairing better since you started tending to it; I knew it would . . . and your wife is paying more attention to you because you applied the same principle at home . . . that's really nice; keep up the good work. If you think a little TLC does a lot for you, imagine the results you'd get from a more concentrated commitment of your time. You can do what you will with your honey at home, but I'd like to see you incorporate these suggestions into your program at work.

Weekly

- Top off cell electrolyte levels and charge batteries
- Repair leaking valves and interconnected piping
- Adjust chemical and blow down bleed rates
- Test safety interlocks per manufacturer's instructions
- Blow out dust from electrical control cabinets and devices
- Observe relays and contactors for arching and chatter
- Operate stand-by units per manufacturer's instructions
- Re-adjust shaft alignments and check couplers
- Check gears and drives for proper mesh
- Clean or replace filters as needed
- Disassemble and clean strainers

- Replace belts and sheaves as needed
- Clean and calibrate combustion igniters

MONTHLY MAINTENANCE

Now that you're displaying a more particular concern for your plant, how is it responding? Outstanding! At home too, you say . . . in what way? Is that so? Good for you . . . the last time my wife cooked me my favorite meal was after my last wage hike. She says people shouldn't be fatter than their pay envelopes, whatever that means. Every month she gets grumpy and says things like that. Coincidentally, every month, my equipment acts up around the same time. Fortunately, I'm able to get both of them back on even keels by applying simple and appropriate logical procedures. In her case that means agreeing with everything she says. I suggest you follow these procedures for the equipment:

Monthly
- Dry out moisture from electric panels and control cabinets
- Megger electric motor and transformer windings
- Inspect and replace machine guards as necessary
- Clean and inspect coils
- Check damper and louvre operation
- Adjust modulating and operating control linkage assemblies
- Adjust belt tensions and align pulleys
- Clean, repair and/or replace heat exchanger fins
- Tighten battery terminals and record specific gravities
- Unclog orifices, flush basins and clean debris from all units
- Check electrical brushes for wear and obstructed movement
- Check gage calibrations and adjust monitoring devices
- Test overspeed trips of all governors in the plant
- Record vibration readings on rotating equipment
- Test safety controls and alarm circuits
- Correct operation of malfunctioning steam traps
- Clear obstructions from spray nozzles and atomizers

- Test circuit breakers for proper operation
- Check relays for dirt, pitting, spacing and alignment of contacts
- Calibrate all diagnostic test devices used for troubleshooting
- Operate backup electrical generators under load conditions

ANNUAL ASSIGNMENTS

It's hard to say what, if any, frequency aspect of maintenance is most important in keeping the plant functioning. Certainly it's all necessary to the operation, but to my way of thinking, annual maintenance is the most critical. Why? Because you can get away with not performing daily procedures now and then; many plants operate through the weekends on automatic controls. And what plant hasn't let preventive maintenance lapse, the weeks their operating engineers were on vacation? Infrequently, non-performance of a month's PM proves of no real consequence as long as the work is performed in the months immediately preceding and following the one which was missed. But there are certain tasks which because of their involvement, cost, necessity or requirement by jurisdictional authority are only performed once a year. If the annual maintenance is not done at that time, then it becomes 2 years between when the task is first and next completed. This increases the possibility of the unit prematurely failing before the next scheduled shut down. Foregoing the annual maintenance on any piece of equipment in your plant is an invitation for Mr. Murphy to pay you a visit. If you don't want to see him in your place, I suggest you incorporate at least the following into your PM program:

Annually
- Disassemble, clean, and inspect equipment interiors
- Replace worn/broken parts on and recalibrate operating controls
- Replace worn, corroded, failed or leaking fluid conduits
- Thoroughly clean and examine electrical panels
- Inspect and replace plant electrical wiring as needed

- Check the condition of all equipment doors and openings
- Replace worn gaskets and seals
- Replace corroded electrical contacts
- Flush equipment interiors and refill with fresh fluids
- Renew pins, triggers, latches, bolts, nuts and fuses
- Renew brushes, set screws, shaft keys and seal rings,
- Renew bushings, cables, springs, clamps & switches
- Renew pen nibs, fittings . . . etc.
- Straighten or replace bent shafts
- Replace broken thermometers, pressure gages and flow-meters.
- Check fans and fan blades for proper operation & integrity
- Dismantle & service all main external operating mechanisms
- Inspect for scoring, cracking, wear, misalignment & erosion,
- Inspect for dirt, tightness, grooving and loose fit
- Inspect for vibration, improper clearance and broken parts,
- Inspect for binding, internal damage, burning and spalling
- Inspect for rubbing, corrosion, overpressure and fouling
- Inspect for deposits, pitting, warpage . . . etc.!
- Hydrostatically test boilers and pressure vessels
- Rebuild, replace or reinforce structural supports
- Replace worn bearings and malfunctioning steam traps
- Replace malfunctioning electric motors and worn controls
- Replace leaking valves and bad sections of piping
- Replace packing in valves and pumps . . . etc.!

PROGRAM ABERRATIONS

Preventive maintenance tasks should be equipment specific and are usually derived from the manufacturer's instructions sheets you receive with your equipment when you take delivery. Tasks are also constructed from the recommendations made by insurance companies, the guidelines of established codes, regulatory body requirements and municipal ordinances. More often than not, however, preventive maintenance programs are established long after the equipment arrives and its associated paperwork has been lost or misfiled by operating engineers having no access to the inspectors, their code books or the law dockets. Subsequently, they must rely on their intuition, experience and reference such as this in order to form a base of information from which to proceed. The preceding tasks and frequencies are normal and customary to the physical plant and are intended as an aid in creating preventive maintenance outlines. Whereas they may be used successfully to that end in a cursory program, they are generic in construction and should not be construed as comprehensive for all equipment.

Preventive maintenance programs need not be limited to caring for physical plant system components and control devices. Any item in the plant can be included that you determine needs periodic attention; such as fire extinguishers, light standards, communications equipment, lightning arresters, vehicle maintenance . . . etc. Nor must you restrict your frequencies to those forementioned in this chapter. Some manufacturers specify exacting work to be performed on their units as a condition of warranty, based on hours of operation. Others claim their equipment is maintenance free, requiring no particular attention to be paid to it. Plant engineers may decide to modify their monthly schedules to 3- or 5-week intervals to accomodate the operating or economic needs of the plant.

The importance of continuing preventive maintenance isn't evidenced by the work performed in caring for equipment or with the timing of its completion; it rests with the knowledge of the operator and his familiarity with the equipment.

Chapter 9

DIAGNOSTIC
EXAMINATIONS

You know it's downright scary how much machines mimic people. Like us they need energy to operate, a regimen of exercise to keep their parts moving smoothly and tender loving care to keep them from getting moody. And like us, they display symptoms resulting from underlying ailments, sometimes just collapsing on the spot; like Suzie over there. Who's Suzie? She's the temperamental sump pump next to number three air compressor; you better switch over to the other pump before we get flooded out. Don't ask. Believe me, you'll be talking with them too after you've spent a few hundred lonely nights here by yourself. Getting back to the discussion; just as people require frequent checkups and/or diagnostic testing, so do our mechanical machines. As an intern, I know this equation still eludes you, so I'll be happy to expound on the truth that it contains. Please take a seat, the doctor is in.

TESTS:

Accumulation

In accordance with American Society of Mechanical Engineers (A.S.M.E.) code rules, a boiler's safety valves must be capable of relieving all the steam that can be generated by the boiler without more than a 6 percent rise above the maximum allowable working pressure (MAWP) or the highest pressure setting of any valve. Accumulation tests can be conducted to make certain the guideline is met. The test is performed by closing all steam outlets on the boiler, overriding the high pressure limit control, firing the unit on high fire and witnessing the pressure rise on the steam pressure gauge. The test should only be performed by competent personnel after the calibration of the pressure gauge has been ascertained and care should be taken to assure that the limit switch and burner modulating controls are properly readjusted before returning the unit to normal service.

Dielectric Absorption

This test is very similar to the insulation resistance test described later herein, but is used specifically to determine the existing condition of a wire's insulation relative to the moisture to which it is exposed. The test is conducted by disconnecting wiring from its normal power supply and substituting a 500-volt direct current potential through it. The voltage is maintained for a set period of time, during which measurements are taken of the insulation's resistance to electrical flow at specific intervals. A graph is then plotted from the values derived from the readings taken and a determination is made of the insulation's electrical integrity. If the graph shows a steady increase in resistance is present, the insulation is considered to be in good condition. If an increase in resistance is not depicted on the graph, current leakage is evident and damp or dirty windings are usually suspected, in which case the unit should be dried, cleaned and retested.

Doble Power Factor

The doble power factor test uses alternating current as a non-destructive means of detecting changes in the strength of electrical insulation materials caused by overheating, ionization or exposure to moisture. Manufacturers use it in their quality control programs to assure the integrity of their products; inspectors utilize it in acceptance testing of electrical devices, and plant engineers incorporate it into their preventive maintenance programs for evaluating the operating efficiency of their equipment.

Eddy Current

A non-destructive test used for detecting deterioration of metal in heat exchanger tubes resulting from erosion, corrosion, pitting, vibration, wear or stress cracking. The process uses a probe inserted into and run the entire length of the tube being surveyed as a magnetic field is applied. When the field is developed, eddy currents are created which are disrupted when passed by defects in the tube metal, registering corresponding electrical signals on the test monitor to which the probe is attached. The signals received are interpreted to determine the severity of the problem, if any, and plans for corrective action are made. Generally speaking, Eddy Current Testing (ECT) is performed by independent laboratories or field service arms of air conditioning equipment manufacturers, employing certified (NDT) non-destructive testing personnel.

Gas Bubble

Everyone past the age of 10 has used or witnessed the use of some variation of this test. Before the tubeless automobile tire was developed, it was the only sure way to find pin-hole leaks in your innertubes. Today, kids use it to test the innertubes of their bicycle tires and bladders in their basketballs. It may be the simplest of tests you might perform, but nevertheless, it's effective. If you want to determine whether any gas-filled object, whose interior pressure is above that of the surrounding atmosphere, is leaking, submerge it in water. Bubbles issuing forth are evidence of a leak whose source can be traced to their origin.

Hydrostatic

Simply put, the term hydrostatic means water pressure; subsequently a hydrostatic test is nothing more than a water pressure test which, in the power plant, is used to determine the pressure integrity of vessels such as boilers, air tanks and heat exchangers. The test is conducted by completely disconnecting the vessel from its system and securing all its openings. Once isolated, all the air is removed from the interior of the vessel as it is filled with water. The vessel is sealed, then using the principle of non-compressability of liquids water within it is quickly pumped up to a predetermined pressure at which it is held during the inspection process; usually, at operating or design pressure or an increased percentage of pressure based on manufacturer's recommendations or codes. After the test, the units are drained and inspected for internal damage before being returned to service.

Infrared

When it is necessary to determine the amount of heat being generated by or contained in a component device, it isn't always possible to simply attach a thermometer to it and read off the temperature. Such is the case with devices which are located in remote areas, are physically inaccessible or contain high voltage electricity that can't be shut down for inspection. That's when infrared or thermal testing procedures are often employed. Infrared is used to locate hot spots by comparing temperatures of areas being studied to the temperature of the area surrounding it or to the ambient air. A camera is used to detect infrared radiation emitted by the object at which it is pointed. The radiation received is then converted to a wider band signal which reproduces the thermal image on a video monitor for study and recording. Plans can then be made to correct the problems found at a later date. In the power plant, the test is performed to check for tightness of electrical connections, blow through of steam traps and heat leak and infiltration studies. Some cameras are capable of providing hard copy black and white photos of conditions found.

Insulation Resistance

The insulating materials used to cover the wiring through which electricity flows can break down or carbonize as the result of exposure to moisture, deposition of foreign matter, current leakage to ground, physical damage or age. To determine the electrical integrity of the covering, insulation resistance tests are conducted whereby the wiring being tested is disconnected from its normal power supply and reconnected to a meggar capable of generating 500 volts of direct current potential. The megger is used to energize circuit wiring for a predetermined period of time under specified conditions of temperature and humidity. The insulation's resistance to electrical conductance is measured in megohms and compared to prior readings to determine whether or not it has deteriorated since last inspected.

Liquid Penetrant

The liquid penetrant test provides for non-destructive examinations of nonporous materials when checking for surface discontinuities such as hair-line cracks on motor shafts, fan blades or pump impellers. A dye is used to coat the surface in the suspect area, then the excess is wiped away leaving it only in the surface openings caused by the flaws. A developer is then applied to the surface enhancing the visual effect to aid the examiner in making a diagnosis. A bright white light should be used when making the examination unless fluorescent penetrants are used in which case the examination should be made under a black light.

Magnetic Particle

Ferromagnetic materials can be checked for cracking on their surfaces or slightly beneath their surfaces using a magnetic particle test. Though similar to the liquid penetrant test, the results are more reliable. There is a dry method, which entails the use of alternating or direct current electricity and a wet method which uses white or black light depending on the type of magnetic particles used. The test consists of magnetizing the area to be examined, then applying dry powdered or fluorescent magnetic particles onto its surface. The particles automatically line up along the magnetic lines of flux created by surface cracks or sub-surface voids providing a visual representation of the fault to the examiner.

Metal Hardness

The strength of metal is measured by its ability to withstand an applied load without failing. The hardness of a metal is indicative of its strength and is measured by its ability to resist indentation when an external force is applied to its surface. Special tests have been developed to determine the hardness levels of materials in which "indenters," made of extremely hard materials, are used to strike the metal under predetermined load conditions causing the metal surface to become indented. The size and depth of the indentation are measured and compared to tables from which the metals "hardness" number or level of resistance is derived.

Overpotential

Overpotential tests are performed to determine whether the insulating coverings of electrical wires are capable of withstanding the loads to which they are or can be subjected. Electricity at a voltage higher than that for which the insulation is designed is passed through the wire it covers. The voltage applied is stepped up in increments, at which time the insulation is checked for current leakage values. Once the required data is collected, the test is discontinued short of causing a breakdown. Though the test can be conducted using either alternating or direct current, direct current overpotential testing is usually chosen because it is less damaging to the insulation.

Overspeed Trip

Overspeed trip mechanisms are safety devices used with constant speed governors to ensure that the rotating equipment they operate don't exceed pre-determined speeds. Testing of the device is accomplished by overriding the governor and allowing the tripping mechanism to shut the unit down at about 110 percent of normal operating speed.

Radiographic

Radiography is used for detecting subsurface abnormalities in metallic fired and unfired pressure vessels and weldment imperfections such as porosity, plate laminations, slag inclusion and lack of penetration. The process entails the passage of short wave radiation, emitted from an x-ray machine or gamma ray-producing isotope, through the material being examined, onto photographic film on the opposite side. Inconsistencies found are represented as light or dark areas on the film depending on their relative density to the metal in which they are contained. Unlike other non-destructive tests, radiographic examination provides a permanent record of the test results obtained for immediate filing and retention.

Safety Valve

Depending on who is charged with setting up the schedule, you might end up testing your safety valves once a day, once a year or never. There is no cut-and-dried rule stipulating how often they need to be tested, and manufacturer's recommendations don't necessarily apply here. The fact is that testing should be done only as often as necessary based on the hours of operation, age and condition of your equipment, the load it normally carries and its operating pressure. You are the judge, jury and jailer on this one. There are two basic means of testing safety valves to assure they will open when your vessel's pressure reaches their setting; one is to raise the manual lifting lever of the valve by hand when the vessel's pressure is within 75 percent of the valve's setting; the second and more desirable method is to override the high pressure limit control and raise the vessel's pressure until it reaches the set pressure of the valve. There is inherent danger in doing it either way . . . so be careful.

Smoke and Dye

Throughout our building's interiors run miles of conduits through which travel wires, fluids and even the air we breath. Much of it is inaccessible, having been buried in the ground or hidden behind walls or in the ceilings. When confronted with diagnosing problems in the dark recesses of these passages, operating engineers often revert to introducing smoke into air ducts or dye into liquid lines, then following the indicator to a point of discharge. The procedure is used for locating ruptures or leaks, determining direction and rate of flow and studying convection currents of fluids.

Soap Bubble

The soap bubble test is the simplest and most underutilized procedure for determining the location of vapor leaks under pressure. It can be used for checking air casing leaks in boiler, leaks from refrigeration lines, gas leaks in fuel tranes and for a myriad of other reasons. The test is performed by generously applying a concentrated soap-and-water solution to the area where a gas leak is suspect, and witnessing whether or not bubbles appear. Bubbles, of course, indicate that leakage is evident.

Steam Trap

The old adage . . . "out of sight, out of mind" holds no more truth than when it's applied to the subject of trap maintenance. On the mechanic's to-do list, steam traps are, at once, the most frequently neglected and costly items to ignore. Malfunctioning traps waste precious energy and raise operating costs. Find yours and test them frequently. The best way is with a test valve hookup attached to the trap body which is used to check for steam blow through. A quick means of determining if a trap is working is to hold a screwdriver blade against it and listen for it to discharge. Another is to check for temperature differences across the trap. Unless your hands are exceedingly calloused, I recommend you wear gloves during the test or perform it with the aid of a pyrometer. Either way, check the incoming steam temperature against the outgoing condensate temperature. If there's a distinct difference between the two, your trap is probably okay; if not, you may want to check it out further for necessity of repair or replacement.

Ultrasonic

Ultrasonic testing can be performed on almost any size and kind of material and is used to detect discontinuities of and below the surface, measure material thickness and find welding flaws. In the process, high-frequency sound waves are introduced into the material being tested which are reflected as echoes from any discontinuities found within it. This is accomplished through the use of a transducer which converts electrical impulses to sound when transmitting and converts echoes to electrical signals for display on an oscilloscope when receiving. By observing variations in the patterns on the oscilloscope, the operator can determine the size and location of the discontinuities he finds. Only one side of the material need be accessible for a complete check of its entire volume.

Vibration

Analyzing the gyrations and pulsations of your equipment beyond their normal operating limits can provide advanced warning of its impending failure. Since all machinery vibrates when operating and given the same loading, any change in its normal rate is indicative of a change in its mechanical condition, then it makes sense to monitor those changes. Vibration analysis is performed by comparing periodically obtained measurements to baseline readings established during the performance of the initial test. Naturally, all measurements are made while the equipment is in operation.

Chapter 10

SPECIAL
OPERATING CONCERNS

There are four major areas, outside of those we've already discussed, that will be of primary concern to you in the operation of your plant. They deal with the combustion process, water analysis and treatment, lubrication requirements and corrosion. Entire volumes have been written on these topics and it would behoove you to include them in your engineer's library for later scrutiny. For now, it will suffice that we address the relevance of the issues. What do you say we jump into water treatment to get us started?

WATER TREATMENT

The Natural Water Cycle

Anyone who's been caught outside during a cloudburst has a good idea where water comes from. The oceans evaporate creating clouds of water vapor which ultimately give up their moisture in the form of rain or snow. A quick and easy explanation, you say; but that's only part of the story. Up until now I was referring to relatively untainted water in its various states. The truth is, water in its natural state is never pure. From the time it begins to precipitate back to earth, it wages a losing battle against contamination. As it's hurled earthward it picks up airborne particulates and gasses. After touching down, it dissolves and absorbs a multitude of constituents as it erodes across the face of the planet on its trek to the sea. And not all of it completes the journey. Some of the water ends up in bogs, lakes, or ponds, some of it replenishes the

ground water we tap into with our wells, and much of it is consumed by animals or vegetable life or evaporates from drying
riverbeds and our own backyards.

Constituents of Water
One thing is certain: by the time it gets to you, the water you
use in your equipment is a veritable slurry of organic and inorganic
compounds and dissolved gasses. Depending on where it is found,
it may be teeming with life or completely devoid of it, as the
result of chemicals added to it by man. The constituents listed
in Figure 10-1 are typical of those found in water samples. As the
plant engineer you are responsible for determining what impurities
are contained in your water, ascertaining what effect they can
have on your operation and implementing a water treatment
program for their control and or removal.

Associated Problems
The most prominent problem associated with the use of
water in equipment is corrosion—in the form of metal deterioration, scale formation—resulting in poor heat transfer and metal
fatigue and fouling, a condition whereby passages, pipes and
nozzles become clogged due to the accumulation of the impurities
deposited within them. No less important, poor-quality water also
results in foaming, whereby contaminated water forms a froth
resembling soap suds in the steam space of boilers. Priming, a consequence of foaming, is the carryover of impurities and water into
steam dischage lines. And embrittlement (a condition of metal
evidenced by hairline cracks which develop in high stress areas
exposed to alkaline salts).

Removing Impurities
On the basis that the water you use at your plant is purchased
from your local sewage treatment facility, it's likely that most of
the suspended matter originally contained within it has been
filtered out using some combination of clarifying processes including sedimentation, coagulation and/or flocculation. How's that?
You're right, it's getting too involved and technical-sounding, but
I'm on a roll, so don't bother me. If you need to know what the
terms mean, look them up in the glossary; that's what it's for. My

Figure 10-1. Water Impurities *(Courtesy: Betz Laboratories)*

Constituent	Chemical Formula	Difficulties Caused	Means of Treatment	Testing Method/Units
TURBIDITY	None—expressed in analysis as units.	imparts unsightly appearance to water. Deposits in water lines, process equipment, etc. Interferes with most process uses.	Coagulation, settling and filtration	Jackson candle (JTU) Nephelometer (NTU) } photometric Formazin (FTU)
HARDNESS	Calcium and magnesium salts expressed as $CaCO_3$.	Chief source of scale in heat exchange equipment, boilers, pipe lines, etc. Forms curds with soap, interferes with dyeing, etc.	Softening. Demineralization. Internal boiler water treatment. Surface-active agents.	Titration—ppm as $CaCO_3$
ALKALINITY	Bicarbonate (HCO_3), carbonate (CO_3)$^=$, and hydrate (OH)$^-$, expressed as $CaCO_3$.	Foaming and carryover of solids with steam. Embrittlement of boiler steel. Bicarbonate and carbonate produce CO_2 in steam, a source of corrosion in condensate lines.	Lime and lime-soda softening Acid treatment. Hydrogen zeolite softening. Demineralization. Dealkalization by anion exchange.	Titration—ppm as $CaCO_3$
FREE MINERAL ACID	H_2SO_4, HCl, etc. expressed as $CaCO_3$.	Corrosion.	Neutralization with alkalies.	Titration—ppm as $CaCO_3$
CARBON DIOXIDE	CO_2	Corrosion in water lines and particularly steam and condensate lines.	Aeration. Deaeration. Neutralization with alkalies.	Titration—ppm as free CO_2
pH	Hydrogen ion concentration defined as: $pH = \log \frac{1}{(H^+)}$	pH varies according to acidic or alkaline solids in water. Most natural waters have a pH of 6.0-8.0.	pH can be increased by alkalies and decreased by acids.	Color comparator pH meter (glass electrode)
SULFATE	$(SO_4)^=$	Adds to solids content of water, but in itself is not usually significant. Combines with calcium to form calcium sulfate scale.	Demineralization.	Colorimetric (photometric) ppm as SO_4
CHLORIDE	Cl^-	Adds to solids content and increases corrosive character of water.	Demineralization.	Titration—ppm as Cl
NITRATE	$(NO_3)^-$	Adds to solids content, but is not usually significant industrially. High concentrations cause methemoglobinemia in infants. Useful for control of boiler metal embrittlement.	Demineralization.	Brucine Cadium Reduction } Colorimetric (photometric) ppm as NO_3
FLUORIDE	F^-	Cause of mottled enamel in teeth. Also used for control of dental decay. Not usually significant industrially.	Adsorption with magnesium hydroxide, calcium phosphate, or bone black. Alum coagulation.	Specific ion electrode. ppm as F
SODIUM	Na^+	Adds to solid content of water. When combined with OH, causes corrosion in boilers under certain conditions.	Demineralization.	Flame spectrophotometry ppm as Na

(continued)

Figure 10-1 (Continued)

Constituent	Chemical Formula	Difficulties Caused	Means of Treatment	Testing Method/Units
SILICA	SiO_2	Scale in boilers and cooling water systems. Insoluble turbine blade deposits due to silica vaporization.	Hot process removal with magnesium salts. Adsorption by highly basic anion exchange resins, in conjunction with demineralization.	Photometric—ppm as SiO_2.
IRON	Fe^{++} (ferrous) Fe_{+++} (ferrous)	Discolors water on precipitation. Source of deposits in water lines, boilers, etc. Interferes with dyeing, tanning, papermaking, etc.	Aeration. Coagulation and filtration. Lime softening. Cation exchange. Contact filtration. Surface-active agents for iron retention.	Colorimetric (photometric) ppm as Fe
MANGANESE	Mn^{++}	Same as iron.	Same as iron.	Same as iron—ppm as Mn
ALUMINUM	Al^{+++}	Usually present as a result of floc carryover from clarifier. Can cause deposits in cooling systems and contribute to complex boiler scales.	Improved clarifier and filter operation.	Colorimetric (photometric) ppm as Al
OXYGEN	O_2	Corrosion of water lines, heat exchange equipment, boilers, return lines, etc.	Deaeration. Sodium sulfite. Corrosion inhibitors.	Winkler method, Rexnord, Chemetrics ppm as O_2
HYDROGEN SULFIDE	H_2S	Cause of "rotten egg" odor. Corrosion.	Aeration. Chlorination. Highly basic anion exchange.	
AMMONIA	NH_3	Corrosion of copper and zinc alloys by formation of complex soluble ion.	Cation exchange with hydrogen zeolite. Chlorination. Deaeration.	Colorimetric (photometric) ppm as N
DISSOLVED SOLIDS	None	"Dissolved Solids" is measure of total amount of dissolved matter, determined by evaporation. High concentrations of dissolved solids are objectionable because of process interference and as a cause of foaming in boilers.	Various softening processes, such lime softening and cation exchange by hydrogen zeolite, will reduce dissolved solids. Demineralization.	Approximation to specific conductance, reported as ppm.
SUSPENDED SOLIDS	None	"Suspended Solids" is the measure of undissolved matter, determined gravimetrically. Suspended solids cause deposits in heat exchange equipment, boilers, water lines, etc.	Subsidence. Filtration, usually preceded by coagulation and settling.	Gravimetric, reported as ppm.
TOTAL SOLIDS	None	"Total Solids" is the sum of dissolved and suspended solids, determined gravimetrically.	See "Dissolved Solids" and "Suspended solids."	Reported as ppm.

intent here is to point out that regardless if the water you purchase from your local authority is aerated, chlorinated, fluoridated or carbonated, it's quality still doesn't meet the strict operating standards of your equipment. Though most of the suspended solids have been removed, the dissolved salts and gasses remain, waiting to be manifested as scale or corrosion. The three major tools used in the physical plant for improving water quality are demineralization (the removal of inorganic salts from solution by ion exchange), deaeration (a process in which dissolved oxygen and carbon dioxide are removed by heating of the water), and internal treatment (a process whereby a variety of chemicals and techniques are used to condition the water to pre-determined values).

Program Benefits

Most organizations, large enough to employ viable operating engineers, take advantage of their professional knowledge to help defray operating costs in the physical plant. Such is the case with water conditioning. Analysis and treatment is an area that can pay handsome dividends if properly performed. The by-products of a well implemented program are increased heat transfer efficiencies, lower fuel expenditures and decreased consumption of chemicals. Except for steam trap maintenance, it has the most potential for reducing annual operating costs in the power plant. If you don't have the expertise in-house, by all means contract it out to a reputable vendor.

Service Agreements

Depending on the magnitude of a company's operation, its policy on contractural arrangements and often times, the political environment, you may or may not be regularly serviced by the people from whom you buy your water treatment chemicals since many companies' purchases consider only the bottom line. The fact is the decision from whom to buy shouldn't be based on cost alone, but on the service you receive from the vendor. Quote me! "Chemicals are chemicals," regardless if they are referred to by their generic nomenclature, their brand name or an alphanumeric designation. Quote me, again! "Any company that provides service is only as reputable as their representative who services your account." Once you find a good one, don't let him get away. The extra money you spend to get and/or keep him will be returned many fold.

CORROSION . . .

Without a doubt, corrosion is one of the biggest headaches facing the operating engineer and the older his plant, the more dedication he'll need for its treatment. I say treatment instead of cure because man hasn't found a way to totally eliminate the problem, only ways of minimizing its effects. In the physical plant, corrosion is the common ground where Father Time, Mother Nature and Mr. Murphy most like to meet as a group. Over time, the metals used to construct your vessels, equipment parts and system components slowly rust in an attempt to revert back to the natural ores from which they were manufactured. The process begins when metal ingots are formed into billets and blooms which are turned into the technological marvels we operate in our spaces. It continues as the carcasses of our aged devices are cast aside onto scrap heaps and only ends when the once-magnificent machines have returned to dust. And while Nature and Time get the corrosion process underway, Murphy just waits in the wings figuring out ways to make your life the most miserable, at the worst possible moments.

The Process

Corrosion is basically an electrochemical process in which electricity flows through a solution of ions between areas of metal. Deterioration occurs when current leaves the negatively charged metal (anode) and travels through the solution to the positively charged metal (cathode) completing an electric circuit, much like the action in the cell of a battery. The anode and cathode can be two different metals or different areas of the same piece of metal. Corrosion occurs when a difference in electrical potential exists between them or as the result of their physical contact. The solution containing the ions is called the electrolyte, which in most cases is essentially comprised of water, its ions and the ions generated by the anode.

Effects of the Process

Although the process remains constant, corrosion is manifested in different ways. The way a particular piece of metal is acted upon depends on its material composition, the temperature and pressures to which it is subjected, the environment in which it is contained and a myriad of other variables, too numerous for consideration in this light dusting of the topic. Generally speaking, types of corrosion are evidenced by the effects of the corrosion process under varying conditions. Pitting, for example, is a localized form of corrosion in which small pockets or voids develop in the surface of metal as the result of the breakdown of protective films or from oxygen concentrations. Uniform corrosion is a general wasting of the metal over large areas of its surface caused by high temperature oxidation, exposure to acidic solutions or containment in corrosive atmospheres. And stress corrosion cracking results from static loading of metal surfaces having built up tensile stresses. Corrosion occurs where there are high levels of oxygen or carbon dioxide, low pH values, where contact is made between dissimilar metals and in damp environments or corrosive atmospheres.

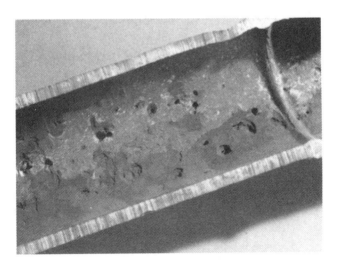

Figure 10-2. Localizing oxygen attack in the form of pitting.
(Courtesy: Betz Laboratories)

Assessing Your Involvement

Don't wait until your plant takes on the appearance of a road-side auto graveyard before deciding you may have a corrosion problem. I submit that the problem began long before your company opened its doors for its first day of business. Corrosion is a constant and unrelenting process that continues until the metals manufactured by man have been returned to the ores from which they were produced. You should check out your equipment before it checks out on you. No metal structure or device should escape your scrutiny. Where should you begin? There is no best starting point; I suggest that you do it in three stages. In the first stage, visually observe the entire plant, detailing which areas show evidence of a corrosion problem. In stage two, open units to internal inspection and review their records of operation, lubrication and water treatment. Finally, send scrapings and samples of corroded parts out for laboratory analysis. The information gathered during the evaluation process can be compiled and studied and a plan of corrective action can then be formulated.

What to Look For

The bulk of your detective work will be accomplished in the second stage of your investigation, as it is the interior metal surfaces of your equipment that are most subject to the ravages of corrosion. The insides of our machines and vessels are frequently exposed to extremes of pressures and temperatures. Many are cooled by or use water in their operation. Their structures expand and contract as they heat up and cool down. And some are adversely stressed as the result of loads imparted on them.

Of the many causes of corrosion, chief among them are oxygen and carbon dioxide liberated from heated water resulting in grooving and pitting. This condition is most evident in the steam and condensate lines of boiler systems. The main culprits in refrigeration units are the calcium and sodium brines used as secondary refrigerants that attack the lines and devices through which the solutions circulate. Improper water treatment can result in poor corrosion control which in turn can cause choking off of narrow passages, nozzles and orifices.

Often times much can be told about the condition of a device by combining a review of its treatment record with an observation of its interior. For instance, if you were to open a fire-tube boiler and found it to be completely clean and devoid of scale, you might consider it to be in excellent condition. But if you were to review the water treatment log for the unit and find that the water contained within it had an average pH value of 6.0 over the past year, you might want to consider whether or not your boiler has been in general condition of uniform corrosion (caused by people on the day shift, of course) and that the vessel may be suffering from a severe loss of metal, subsequently lowering its pressure integrity.

There are a number of critical things to look for when checking for corrosion in your plant, but as you can see from the last example, probably none more important than good, sound advice from an expert. Any time you open a piece of equipment in your plant, it's always a good idea to have an inspector or manufacturer's representative there to provide you with his opinion of any problem found.

Limiting Your Exposure

Obviously, the best way of defensing your equipment's insides against the damage caused by corrosion is by properly filtering, treating and deaerating the liquids you inject into them. But what about the exterior of those units? You can limit your problems by shopping around for units manufactured from corrosion-resistant metals or substituting non-metallic-constructed units for traditional metal ones when purchasing new; or you can accept the fact that corrosion will always be with us and draw up a plan for managing it. Depending on how much you estimate your potential savings might be, you may or may not want to have a consultant hired to study your plant and design a program, but in any case you should make corrosion prevention and control a priority in your operation. Some things that you can do to lower the incidence and severity of corrosion are:

- lower the operating temperatures of moving parts
- substitute corrosion-resistent materials
- limit loading stresses on your equipment
- use only high-quality stable lubricants
- keep different metals from physically touching

- strictly adhere to all PM guidelines
- maintain water quality within recommended parameters
- keep all connections tightened and repair corroded areas
- filter out contaminants in equipment fluids
- don't overtighten flange and connection bolts
- eliminate sources of vibration
- maintain pH levels within set limits
- limit the conductivity potential of plant waters
- deaerate feedwater before introduction into your boilers
- maintain proper stack temperatures
- use cathodic inhibitors in your treatment program
- lower relative humidity in the plant
- redirect corrosive fumes away from equipment
- keep spare parts packed in grease until used
- lower the speed at which liquids are transported
- use protective primers on all metal surfaces
- repair leaks and replace bad insulation
- use neutralizing and filming amines in water systems
- keep non-metallic liners in metal tanks in good repair

COMBUSTION . . .
The Mechanics of Burning

Have you ever seen someone throw a lighted cigarette into a bucket filled with gasoline? I have. No, it didn't explode; it didn't even catch fire. As a matter of fact the gasoline put the cigarette out. Please don't attempt this yourself because if it's done improperly the gasoline can catch fire, very easily! So why didn't a fire start when I witnessed the event? Because something was missing. There are two basic premises in combustion theory:

- combustion only takes place on a vapor level
- a specific ratio of fuel vapor and oxygen, at the proper ignition temperature must exist for combustion to occur

So what was missing? Perhaps there wasn't enough vapor being generated or the heat from the cigarette wasn't in contact with the vapor long enough. I'll never know for sure. You see, the experiment wasn't performed in a laboratory under controlled conditions by a qualified combustion expert. It was done at a local service station, near the fuel pumps, by a 16-year-old whose mental prowess was suspect. Not that he couldn't get away with extinguishing a cigarette in this manner time after time, because he did, but I contend that he was lucky . . . very lucky, to tempt fate and win as he did. What if the outdoor temperature were higher that day and the volume of vapors given off by the fuel were substantially increased? What if the tip of the cigarette broke off and its ember were atomized as it passed through the volatile vapors? What if the wind shifted as the ember passed through the vapors. The vapor theory is the reason why candles can't burn without the aid of a mechanical wick; different size grates and combustion chambers are specified for high- and low-volatile solid fuels, and why fire extinguishers are so effective at extinguishing large fires when their contents are discharged at the base of the flame.

Figure 10-3. Fuel Oil Tank

Fuel in The Plant

Outside of manufacturing industries that use exorbitant quantities in their production processes, Stationary Engineers are probably the largest commercial consumers of fuel in North America. They burn every type imaginable and many have plants that are veritable fuel depots. Tanks containing propane and butane dot the landscape of facilities that don't have access to natural gas pipelines. Fuel oil tanks are located above or buried below ground for supplying boiler furnaces when their normal gas supply is interrupted. Diesel fuel is stored for use in emergency generators used to provide electricity when utility power is lost. Gasoline pumps supply fuel for operating plant vehicles. Plants that use solid fuels, such as coal, wood and even refuse, stockpile their supply in bins, often maintaining reserves of several months. In addition to the autonomous fuel banks they may possess, most plants purchase fuel from their local utility companies. Natural gas is piped into the plant to fire boilers, domestic hot water tanks, space heaters and dietary food processing equipment. Electricity, though usually not considered a fuel itself, is used for many of the same purposes as well as for resistance heating of fuels, baseboard comfort heating and illumination.

Characteristics of Fuels

Fuels come in three physical states—solid, liquid and gas. As the combustion process is the rapid oxidation or combining of fuel atoms with oxygen atoms, fuels must first be reduced to their vapor form before they can be burned. Gaseous fuels readily combine with oxygen and need only be brought up to ignition temperature. Properly oxygenated, they are generally clean burning, leaving little or no residue. Approximate heating values of gases range from 1000 to 3000 BTU's per cubic foot. Liquid fuels must be transformed to their gaseous state before being burned but once vaporized, generally share the same requirements as those for gaseous fuels. The portion of the liquid fuel that is not vaporized is evidenced by the carbon it deposits throughout the combustion chamber and gas passes of the unit in which it is burned. The principal liquid fuel is oil which is referred to by grade as light, number 1 and 2; medium, number 3 and 4; or heavy, number 5 and 6. Approximate heating values range from

135,000 to 156,000 BTU's per gallon. Solid fuels must be heated to release the volatile materials contained within them. They are the most difficult fuels to prepare for the combustion process and leave behind large quantities of ash when burned. The principal solid fuel is coal, but anything that burns can be used as a fuel if your equipment is designed to handle it. Approximate heating values of coal range from 7,000 to 15,000 BTU's per pound.

Figure 10-4. Propane Tank

Combustion Theory

Up to this point we have referred to the combustion process as rapid oxidation of a fuel source. Whereas that is a true summary of what occurs during the fuel-burning process, it is also a misleading statement because pure oxygen isn't generally available, nor is it desirable for combining with fuels in the physical plant. It may come as a surprise to you but oxygen itself is not flammable; it only supports combustion. And the purer the oxygen is that's combined with a fuel, the hotter and faster the fuel will burn. As the purity of oxygen and its ratio to fuel increases, it ceases to be a catalyst for combustion and becomes an ingredient of an explosive mixture. So it's just as well that the oxygen we use for burning our fuels in the physical plant is derived from the same diluted air we breath which, by weight, contains about 21 percent oxygen. During combustion, the oxygen derived from the air combines with fuel vapors resulting in the liberation of thermal energy. To what extent fuel is consumed is dependent on several factors. If perfect combustion is your objective, chances are you'll be disappointed with the results. Theoretically, PERFECT COMBUSTION can only occur when each atom of fuel is brought into contact with the exact amount of oxygen needed for its consumption at a precise temperature and held there long enough to complete the process. But if your objective is simply to complete the combustion process, your chances are better. COMPLETE COMBUSTION is accomplished by supplying more than the theoretical amount of oxygen into the fuel mixture to assure full consumption of the fuel. In boiler furnaces, this is accomplished by supplying "primary" combustion air through air registers at the burner, then introducing "secondary" air to aid in mixing the air with the fuel. This results in the total consumption of the fuel, but some oxygen is always evident in the exhaust gasses. The higher volume of air supplied is referred to as "excess" air and is a function of the draft system. INCOMPLETE COMBUSTION happens when one leg of the fire triangle is affected; either the fuel mixture is too "rich" or too "lean," it hasn't reached ignition temperature or it isn't sustained for a long enough period of time.

LUBRICATION . . .

Whether it's carried out by the operating engineers, preventive maintenance mechanics or persons designated as oilers whose sole responsibility is to keep their machine's moving parts moving smoothly, a properly implemented equipment lubrication program is essential to the economical operation of a power plant. And here are some things he should know before he grabs his oil can and indiscriminately starts squirting.

Its Basic Functions

So why are we always oiling our machines? The answer seems simple enough: if we don't, they'll gradually bind up and eventually stop running altogether. But the need for lubrication can't be that easily explained away because there's more to it than just making metal surfaces slippery. Aside from reducing friction, lubricants perform in a variety of other ways. They are used to flush out contaminants that accumulate in oil passages and reservoirs. When applied to metal they inhibit the corrosion process. Trapped in the mesh area of gear teeth they provide a cushioning or shock absorbing effect. Lubricants also provide a cooling effect on equipment, limiting the stress imposed by heat on their metal parts. They also serve to seal out dirt and other foreign matter from internal working parts and depending on the application, are even used to transmit power hydraulically. So you see, there's much more to lubrication than packing a zirk fitting with grease or filling up an oil cup.

Types of Lubricants

Lubricants are classified by state as liquid, semi-solid, solid or gaseous. Liquid lubricants are derived from hard (stearin) or soft (lard) animal or fish fats; vegetable plants and seeds such as cotton and soybeans; refined from crude as mineral oil, or synthetically produced by man using non-petroleum-based substances in conjunction with some combination of these. Oils derived from petroleum are either asphaltic based, containing heavy tar-like materials; paraffin based, possessing large amounts of wax-like materials; napthenic based, containing large amounts of naptha; or mixed-base, bearing some or all of these components. When thickeners (soaps) are added to oils, they become semi-solid greases. Greases range from hard to soft and are recognized by their pumpability, resistance to dissolving in water, ability to maintain their consistency and their melting points. Solid lubricants are formed metallic or chemical compounds used in applications at temperatures outside of the effective operating ranges of oils and greases. Like solid lubricants, lubricating gasses are used in circumstances that would render the other types ineffectual.

Properties of Oil

Aside from color, thickness and smell, there's no reliable way to distinguish one oil from another, depending only on the human senses to evaluate them. But there are important distinctions that must be considered when matching an oil to its intended application. How hot will it get? Will it be exposed to moisture? How much air (oxygen) will it contact? These questions, and many more, must be answered in order to make the proper choice. The most commonly considered properties are the oil's flash point (temperature at which its vapor will sustain a flame), fire point (temperature at which its vapor will sustain a flame), viscosity (how thick it is), and pour point (temperature at which it ceases to flow). These properties, and many others, are often engineered into the oil to meet pre-determined specifications through the use of additives which enhance their performance. A few such additives are antifoam agents which break up air bubbles that form when oil is circulated, detergents that hold dirt particles in suspension and keep them from depositing on metal surfaces, and antioxidants which prevent acid formation resulting from the oxidation of oils. Additives are also used to increase an oil's slipperyness, its ability to remain separated from water and to reduce the effect of temperature on its viscosity. The fact is, oil can be treated to meet almost any requirement.

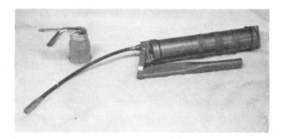

Figure 10-5. Lubricators

Plant Applications

Knowing your equipment's operating constraints is critical to the selection of lubricants. Operating speeds, temperatures and pressures, hours of operation, loading and even environmental conditions need to be taken into account. Refrigeration systems, for example, operate over wide temperature ranges and require wax- and moisture-free oils to avoid problems of ice and acid formation. Electric motors come in sizes from fractional to thousands of horsepower, and have lubrication requirements running the gamut, from good quality machine oils to heavy greases. Gear oils are used in applications where film strength is a determining factor when considering the pressure or loading involved in mechanical power transmission. Where the speed of rotating parts is of prime importance, low viscosity spindle oils are often specified. If you have any doubt as to which lubricant is appropriate for a particular plant application, it's best to contact the manufacturer for a recommendation.

Lubricant Storage

Most physical plants don't require extremely large quantities of lubricants to be maintained for their daily use, but due to the diversity of their operations they often stock many different kinds. And as most oil and grease is packaged in units that are usually not completely consumed in one application, they pose a special storage problem for the operating engineer. Add to that the slippery nature of the products and the problem becomes a hazard. Compound that with their flammability and the hazard becomes an accident waiting to happen. But accidents need not occur. All that's necessary to avoid the downside-of-the-lubricant-equals-disaster equation is to use a little common sense during their handling and storage. Here are a few thoughts on the subject you might wish to consider when evaluating your plant's procedures:

- store all lubricants in well ventilated areas
- supply dry filtered air to storage rooms
- install fire protection equipment in storage rooms
- keep all open containers tightly covered
- label the contents of all open containers
- discard old lubricants in approved containers
- never use contaminated lubricants
- electrically ground all metal drums
- store all lubricants safely above the floor
- never leave oily rags laying about
- keep lubricants separate from potential contaminants

- clean up oil spills when they happen
- maintain lubricant accessories in good repair
- use proper techniques when handling containers
- keep sharp objects away from containers
- allow no open flames in storage rooms
- discard empty and damaged containers
- protect lubricants against moisture exposure
- maintain storerooms within recommended temperatures
- keep records of lubricants stored and used

Chapter 11

STAYING
OUT OF TROUBLE

Well, my friend, you've come a long way. I've seen you err and recover, completely lose control of a situation and at times, display flashes of brilliance in your work. We've learned a lot from one another and I'm going to miss you. But tomorrow you'll be on your own and I won't be here to bail you out if you get into trouble. So that you won't self destruct in my absence, I'll leave you with these pearls of wisdom to comtemplate. They were derived from the accumulated knowledge and experiences of many engineers who preceded you. Heed their messages.

Safety

Safety. What does it mean? How important a consideration is it? How much money should be spent on its pursuit? These are tough questions whose answers are as varied as the opinions of the people and situations that ask and prompt them. Webster's Dictionary defines safety as "the condition of being safe from, undergoing or causing hurt, injury or loss." In the physical plant, that covers a lot of ground. Not that I consider myself more capable than the G & C Merriam Company of describing the term, but as an operating engineer, here's what it means to me. "Safety is a state of mind which is manifested in the actions taken by the persons charged with securing it." Now it may seem to you that these two expressions run the gamut from the simple to the sublime, but I've got a good reason for coupling them here.

On the surface it might appear that establishing and maintaining a safe physical plant is simply a matter of using your God-given common sense when planning and performing your daily tasks. Whereas that's a large part of the process, there's a lot more to it than just using your noodle while you work. Safety is a multi-faceted consideration, entangling every aspect of physical plant construction and operation with the decision-making process. It involves the structural stability of the buildings housing your equipment as well as the performance parameters of that equipment, details how hazardous materials must be handled and stored and dictates the behaviour of people under special environmental circumstances. It impacts on plant operating costs, employee health and the economies of surrounding communities. The safety issue poses many problems and concerns. It's often an elusive subject when considering the expense of its implementation, but always an expensive proposition when it's not considered. Aside from moral and ethical implications; engineering principle, regulatory statutes and available funding all play important roles in the safety scenario. To what extent each bears weight on the issue depends on the problem, political environment and personalities involved.

Safety Programs

Whether you're naturally safety conscious, or you've acquired the trait by order of your organization, is irrelevant. The fact is, if you're not, you or your company will eventually suffer as a result. And that's only half of the story. Regardless of how aware of or concerned about safety a company or its employees may be, a totally safe operation can never be guaranteed. The best that can be hoped for is to achieve some degree of control over a plant's losses by minimizing its exposure to risk. This can be accomplished through the implementation of comprehensive safety programs covering all aspects of its operation whether inanimate, mechanical or human in origin. Every organization is unique and fashions its programs based on individual need. It would behoove you to acquaint yourself with yours, paying particular attention to procedures involving the operation of the physical plant. Safety programs come in many variations and complexities, but generally speaking address the issue under the headings of general safety, electrical safety, fire safety, security and emergency operations.

Operating Precautions

Over the long haul, safety programs are our most effective means for controlling the incidence and severity of physical damage and personal injury in the plant, but it's through the use of common sense in everyday situations that we derive the most immediate benefit. Constant, thoughtful vigilance results in a relatively problem-free operation. On the other hand any lapse in concentration invites trouble to rear its ugly head. Here are 100 don't and do's to keep it from jumping out at you.

Never

- exceed allowable working pressures
- block combustion air supplies
- permanently override a control setting
- hammer parts under pressure
- use pipes to increase the leverage of wrenches
- place combustibles near hot surfaces
- fill a hot, empty boiler with cold water
- try to vary the setting of a safety valve
- permanently override a speed-limiting governor
- smoke in combustibles storage areas
- leave electrical wires exposed
- permanently by-pass a safety interlock
- raise the temperature of fuel above its flash point
- use tools that aren't designed for the work
- operate a malfunctioning unit
- jury rig an operating control
- cap or plug exhaust piping
- use a cast iron boiler as an incinerator
- paint over the data stamped into nameplate
- enter a fume-filled compartment without purging it
- try to turn energized equipment by hand
- look directly into an air stream without goggles
- wear loose clothing around rotating machinery

- use equipment for other than its intended purposes
- operate machines that are out of balance
- lose your temper in a power plant
- overlubricate or underlubricate equipment
- push or wrestle with people around equipment
- allow alcohol to be consumed on the premises
- permit unauthorized personnel in the plant
- enter boiler drums without a partner outside
- spray water onto hot brickwork
- install a device not suited for the application
- overtighten connecting bolts and stud nuts
- allow nozzles and sprayer plates to become clogged
- start a cold boiler on high fire
- start up equipment without first inspecting it
- remove fuses or breakers with your bare hands
- work on a machine you're unfamiliar with
- use broken tools to fix broken equipment
- straddle wires or small pipes with ladders
- stand in front of pressure-discharging ports
- open or close valves suddenly
- fire superheaters with the inlet valve closed
- diagnose problems with uncalibrated equipment
- work on energized electrical circuits
- allow distractions from radios, televisions . . . etc.
- jury rig bridges on high surfaces
- indiscriminately mix chemicals or fuels
- take action when in doubt of the result

Always

- test controls and devices regularly
- keep gage glasses clean and free flowing
- repair leaks immediately

- handle hoses carefully when wetting down
- pile hot ashes away from equipment
- report personal injuries immediately
- shut down when in doubt
- allow welded parts to cool naturally
- keep two nuts on flanges when separating them
- clean up spills immediately
- keep machine guards on
- shut down equipment before making repairs
- vent rooms where combustibles are stored
- use proper techniques for lifting heavy objects
- remove empty containers from the work area
- unplug power tools before leaving an area
- relieve pressure before breaking joints
- allow equipment to cool naturally
- check units for proper combustion
- purge furnaces with fresh air before lighting
- replace malfunctioning thermometers
- use explosion-proof drop lights
- inventory your tools after making a repair
- use gloves when handling hot items
- make certain room exhaust fans work properly
- remain clam and in control at all times
- keep equipment surfaces clean and dry
- replace warped or cracked parts
- let relief personnel know about changes you've made
- maintain proper fluid levels, pressures & temperatures
- repair the cause before pushing the reset
- repair broken windows in cold climates
- comply with chemical company recommendations
- lay up units per manufacturer's instructions
- keep expansion and air tanks drained

- replace worn or burned electrical contacts
- record equipment performance problems
- read the engineer's log before starting work
- wear the clothing that the job calls for
- keep spaces containing batteries well vented
- check fire extinguishers for proper charge
- repair torn lagging and insulation
- make sure that alarms and indicator lamps are working
- maintain proper air-fuel ratios
- keep fuel oils and lubricants free of contaminants
- tag units "out of service" before making repairs
- keep supplies and materials properly stored
- perform preventive maintenance duties in a timely fashion
- adhere to company policy when handling emergencies
- stay alert and respond immediately to problems

THE INSPECTION PROCESS

In the physical plant there are more ways for getting into hot water than just ignoring safety rules or underutilizing your gray matter. If you don't check out your equipment from time to time, small problems which could have been caught during the inspection process can burn you later. Since machines and vessels come in multiple sizes and configurations, are subjected to a wide range of operating conditions, and are used for myriad purposes in differing environments, it stands to reason that no one inspection procedure can be applied to all the equipment we maintain. Subsequently, I've decided to take a generic approach to familiarizing you with the process. You'll find that the following procedural steps can be applied to most equipment found in the physical plant:

Scheduling and Notification

- alert all end users to the planned outage
- pick a time which best suits the overall operation
- notify regulatory personnel of inspection dates and times

Equipping for Inspection

- assemble the following:
  - first aid kits, cots and blankets
  - low-voltage, explosion-proof lights
  - inspection mirrors and diagnostic devices
  - pad locks, chains and out-of-service tags
  - appropriate hand and power tools
  - ladders, scaffolding and hoisting gear
  - vacuum and compressed air sources
  - safety gear and back-up man
  - solvents and lint-free rags

Room Preparation

- ensure the room is well lit and uncluttered
- provide a work bench for dismantling and cleaning
- maintain constant and proper ventilation rates
- provide a clean area for taking breaks and lunches
- install eye wash, shower and first air stations

Preparing The Unit

- allow equipment to operate through its final cycle
- isolate all sources of fuel feed
- relieve all pressure from the unit
- shut down and tag all auxiliaries
- disconnect cross connections to other units
- drain all appropriate fluids
- open access doors and inspection ports
- physically clean the unit's interior
- remove and dismantle all appropriate appurtenances

Internal Inspection

- check for the following:
  - loose, worn, or missing parts

- corrosion in the form of pitting or grooving
- misalignment of nozzles, flames or shafts
- sagging, bulging or blistering of metal parts
- bad weldments and structural infirmities
- fouling and deposition in pipes and passages
- evidence of cutting action or erosion
- warped seats and valve stems
- burning and scoring of metal surfaces
- proper sizes and settings of valves and controls
- correct piping arrangements
- stress cracking in tube ligaments
- broken bolts, stays . . . etc.
- deterioration or wiring and insulation
- cleanliness of control device interiors
- broken baffle plates and blocked air passages

External Operating Inspections

- check for the following:
  - proper range of motion of linkages
  - adequate supply of combustion air
  - appropriate air-fuel ratios
  - functionality of operating controls and safety devices
  - dependability of control limit settings
  - adequate pressure relief capacities
  - proper venting of exhaust gasses
  - safe points of overpressure discharge
  - proper lubrication of all moving parts
  - correct operating pressures and temperatures
  - broken gage and inspection glasses
  - obvious deformities of exterior surfaces
  - evidence of cracking or burning of metal parts
  - leaks under insulated surfaces

> — sustainability of proper operating levels
> — proper operation of auxiliary devices
> — correct piping configurations
> — proper wire and fuse sizes
> — loose connections and fittings

Finish Work

- review and discuss the inspector's findings
- log all pertinent data into the record
- repair all discrepancies found
- renew gaskets, glasses and fluids
- reorder parts used from stock

ODDS AND ENDS

If I were put upon to name the most prevalent cause of trouble in the plant operations field, I would have to say that it's ignorance. An operator can be given a comprehensive set of instructions detailing strict procedures to be used for ensuring safety in the plant and proper maintenance of its equipment, but unless he has the knowledge necessary to carry them out, he'll eventually run into trouble. Knowledge, in that sense of the word, isn't just the understanding of what things are and how they work; it's having a ken for when things aren't working properly and knowing what to do about it. Power plants are fraught with unusual noises and mysterious occurrences which even seasoned professionals have difficulty deciphering and explaining. In order to deal with these peculiarities, an operator must have a high mechanical aptitude, a keen insight into the physical laws of nature and be totally familiar with his equipment, its operation and the structures which house it. To save you from straining your brain, I've put together this potpourri for your edification.

Water Hammer

If your water or steam piping bangs and clangs or you hear rumblings in them that end in a loud thud vibrating your entire building, you've got a water hammer problem. The most common cause of this condition is the sudden closing of valves or faucets or the lack of surge piping in the circulating loop of domestic water systems. Other, more damaging causes, are condensation of steam in a closed pipe, carryover of boiler water into steam discharge lines and the sudden starting or stopping of pumps. Water hammer problems can be reduced or eliminated by avoiding the installation of downward dips in steam lines, installing "dead" pipes in domestic water lines, bleeding air out of water systems, keeping water out of steam systems, filling tanks from the bottom to prevent air entrapment, properly pitching condensate lines and through the use of pressure-reducing valves.

Cavitation

The American Heritage dictionary defines cavitation as "the sudden formation and collapse of low-pressure bubbles in liquids by means of mechanical forces, as those resulting from rotation of a marine propeller." Although the condition can also exist at valves and reducing stations, I define it as "a pain in the ass inflicted by a pump designer's pencil." If you can relate the impeller of a centrifugal pump to the marine propeller mentioned above, your starting to catch my drift. Cavitation affects discharge pressures, is a catalyst for erosion, contributes to water hammer problems and causes damage to valve discs and seats. The best bet for minimizing the incidence of cavitation in pumps already installed is to provide a recirculation line to a deaerator feeding the pump. The percentage of recirculation required increases and decreases proportionally with the temperatures and pressures of the liquid being pumped.

Low Water

Continued operation of water levels which have dropped below normal is the major cause of damage to boilers and hot water heaters. Low water conditions can result in bulging of metal surfaces, ruptured tubes, steam explosions, personnel injury and loss of life. Factors that contribute to the condition are improper blow down procedures, defective low water fuel cut-offs, improper admission of feed water, jury rigged or by-passed operating controls and operator error. When it becomes evident that your unit is firing while suffering a low water condition, you should immediately close off the fuel supply and allow the unit to cool, naturally. *NEVER ADMIT WATER TO A UNIT WHILE IT'S IN A LOW WATER CONDITION, AS THE WATER WILL QUICKLY FLASH TO STEAM WHEN IT CONTACTS THE HOT METAL SURFACES, THEREBY CREATING A POTENTIAL FOR EXPLOSION.* Low water conditions can be prevented by using proper blow down techniques, verifying gage glass indications, frequently testing low water fuel cut-offs, replacing defective controls and paying regular attention to components in the feed water trane.

High Water

Though it may not culminate in the same dire consequences as operating with low water levels might, a high water condition in your boiler can pose some significant problems of its own. Operating with water exceeding normal levels can result in damage to steam lines, their component parts and the equipment to which they are attached. No less important are the underlying causes of high water which include faulty feed water regulators, plugged gage glasses and water column connections, problems in the feed water system and contributing factors such as foaming and priming. Immediate relief of a high water condition is accomplished by partial drainage of the unit through blow down lines after which the cause should be found and corrected. *STEAM BOILERS SHOULD NOT BE FIRED WHEN HIGH WATER PROBLEMS EXIST.*

Explosions

An explosion is a sudden, rapid, violent release of mechanical or chemical energy from a confined space. In the power plant they occur mainly in two ways: tearing asunder of vessels due to overpressure conditions, or as the result of extremely rapid oxidation of combustibles. Now I know the purist will ask you why I didn't cover the explosions of fly wheels caused by mechanical forces and he'll be right; that's a type of explosion you'll find in a power plant, but then I also didn't mention nuclear fission. For obvious reasons, those subjects involve more in-depth consideration. For now we just need to touch on the basic ways you're most likely to get into trouble. After all, there's no need to blow the topic out of proportion.

When a vessel is in an overpressure condition, that can mean two things: either the interior surfaces of the vessel are being subjected to more pressure than what it was designed for, or more than it can withstand in its present condition, even if that value is lower. Overpressure explosions result in catastrophic physical damage, major injury of personnel and death. Their causes are innumerable and include deliberate gagging of or defective safety valves; low water conditions in boilers and hot water heaters; scaled, corroded or thinned surfaces; and contributing factors such as defective pressure gages, by-passing or tampering with controls and poor maintenance. These types of explosions can be avoided by incorporating proper corrective and preventive maintenance techniques, frequent inspection of operating controls, following manufacturer's guidelines for testing safety interlocks and simply considering the consequences of your actions.

Combustion-based explosions also cause loss of life and property damage—often in greater numbers and severity than overpressure explosions—due to the fire and smoke which accompanies the flying debris. That's not to say that fire and smoke are never evident at the site of an explosion caused by an overpressure condition, because they often are. The major difference between the two types is that fire forms a part of and is concomitant with the combustion-based explosion and only infrequently follows or is consequential to the overpressure-based explosions.

Among other items on an endless list, fire-based explosions result from poor fuel handling and storage practices, improper air-fuel ratios, defective operating controls, by-passed safety interlocks, leaking fuel lines and poor maintenance practices. Prevention is best accomplished by following strict fuel handling and storage guidelines, maintaining adequate volumes of combustion air, providing meticulous care of fuel-burning equipment, controls and devices, closely monitoring the various combustion processes ongoing in the plant and keeping constantly on the alert for anything out of the ordinary.

Safety Valves

The terms safety valve, relief valve and safety relief valve are often used interchangeably and therefore incorrectly. Though all three are automatic pressure-relieving devices actuated by upstream static pressure, each is used for a specific application. *Safety valves* are used for steam, gas or vapor service and are characterized by their rapid full opening or "pop" action. *Relief valves* are used primarily for liquid service and open in proportion to the increase in pressure over the opening pressure. *Safety-relief valves* are suitable for use as either safety or relief valves. A rule of thumb that can be used to distinguish between liquid and vapor service valves is to read their nameplates; the relieving capacity is listed in pounds per hour on safety valves and in Btu's per hour on relief valves.

But my intent here isn't to show the differences that exist between the types; I only mentioned them to make you aware that there are some. Actually I wanted to touch on these similarities.

They should:

- provide positive, automatic action
- be located at the highest point on the vessel
- be piped to a safe point of discharge
- be attached as close to the vessel as possible
- be tested frequently per the manufacturer
- never be gagged, by-passed or removed
- have no intervening valves in their lines
- be installed in the correct quantities
- have their settings adjusted only at the factory

A Final Word

As you can see, I didn't put this chapter together with any particular rhyme and for a definite reason. When trouble occurs in the plant, it doesn't happen by category, in alphabetical order, or to a schedule. Trouble comes in many forms, is extremely difficult to predict and impossible to defense against 100% of the time. It causes lost time, wastes energy and costs money. The best that you can hope to do is to hold your own in the struggle. When it confronts you, maintain a level head and remember what you've learned. Unless you have any questions, that's a wrap. How's that? Me too. I'll send you a card from the islands.

APPENDIX A

National Institute for the Uniform Licensing of Power Engineers, Inc.

(Incorporated November 22, 1972)

PREFACE

THROUGH THE YEARS Power Engineering has meant different things to that somewhat diversified group practicing in this field. The Power Engineer in the northern latitudes will, of course, be more involved in heating and steam generation than his more southerly counterpart, where refrigeration may be the principal objective. Likewise the originating energy source will vary with the geographical location and operational requirements to include oil, coal, hydraulics, gas, etc.

As our society becomes increasingly mobile, we find common denominators being sought and implemented in numerous areas never before invaded; the Federal Government is stepping into regions heretofore sacred to the several states and in some instances to the municipalities therein.

The common denominator so necessary in many fields of engineering, including the field of power engineering, that best equates to and makes the equation of cognizance a solvable entity and useful tool is that of professional standards.

With our mobile society it becomes necessary to develop a standard set of qualifications in the field of Power Engineering to insure the safe operation of all types of power-generating equipment.

No man can be expected to become expert in all phases of Power Engineering through his experience. It is, however, important that he have the necessary exposure to recognize and search out that expertise when the occasion demands. He must constantly update his library, his codes, his standards and his knowledge.

One of the purposes of these standards is the encouragement of continuing education of power engineers in their chosen profession; to update the man and his reference library; to keep abreast of the never-ending development of newer, better and safer technological innovations.

No Power Engineer training and licensing program can be expected to totally remove the possibility of power plant accidents any more than a driver licensing program has been able to render our highways free of automobile accidents.

Let us, in the promotion of these standards, endeavor to stamp out the factor of ignorance largely responsible for accidents in the field of Power Engineering.

James F. Schmid
Peter H. Burno

ITS PURPOSE

The National Institute is a third-party licensing agency that acts on a national level for power engineers and those associated with the profession. (Registration is purely voluntary on the part of licensing agencies, etc.) It is, however, the only formal structure through which a person in power engineering may establish, formally, a level of national competence and national professional recognition. NIULPE is affiliated with the National Association of Power Engineers and will perform the following major functions:

1. Establish and maintain uniform standards as to qualifications for power engineers.

2. Promote safety as per preamble.

3. Assist licensing agencies to determine the competence of power engineers through investigations and examinations which test the qualifications of voluntary candidates for lincenses to be issued under the rules of the Institute.

4. Grant and issue commissions to licensing agencies applying and qualifying.

5. Maintain a registry of all licensing agencies meeting the requirements of NIULPE.

6. Serve as a clearing house to facilitate reciprocity between all licensing agencies.

7. Maintain a registry of all examiners who meet the requirements of NIULPE and are commissioned by NIULPE.

8. Develop and keep current uniform terms and definitions, updating terminology and eliminating obsolete terms.

9. Encourage the enforcement and compliance to all codes, laws, and acts assuring the protection of health, life and property, (i.e., Occupational Safety and Health Act of 1970, Public Law 91-596, and the accepted engineering practice standards in appendices B and C in the Boca Basic Mechanical Code/1971, Air Pollution Control, Article II of the Boca Mechanical Code 1971).

10. Publish a standard curriculum for the study, education, requirements, power equipment rating, automation, etc.

11. Act as advisor to educational entities engaged in teaching of power engineering and technology.

THE POWER ENGINEER

A power engineer shall be defined as one skilled in the management of energy conversion. A power engineer operates and maintains equipment essential to power generation, heating, ventilation, humidity control, and air conditioning in industrial plants, institutions and other building complexes. He performs in a responsible manner as a technical expert in operating, maintaining and repairing engineering plants consisting of steam boilers, pressure vessels, internal combustion engines, steam engines, turbines, refrigeration and air conditioning equipment, motors, pumps, compressors, distillation units and similar equipment.

He must operate and maintain equipment according to state and local laws and codes, because the health and safety of many people depend upon the proper operating and functioning of the engineering equipment. As a lower-grade power engineer he is under the general direction of a chief or engineer holding a higher-grade license whose license is equal to or above the requirements needed to operate the plant, and perform those duties prescribed by the higher-grade engineer. As he advances his licenses through experience and study, he becomes qualified to take charge of an entire engineering plant operation. Through on-the-job training and application of study courses, he attains the skills and knowledge required.

A power engineer may advance to many jobs associated with the profession as he becomes adequately educated and trained. Under proper engineering direction, he may perform duties of an engineering technician, draftsman, or estimator, designing, planning, estimating, erecting, inspecting and testing engineering equipment, servicing and testing of materials, sales engineering and representation, or instructional activities.

When licensed, the power engineer has clearly established a level of competency. As his skills are advanced by experience and education, he has the opportunity to improve his level of certification through the five license classifications. Encouragement toward professional advancement is a vital element toward sound personnel policies. Advancement in licensing should move the power engineer closer to the management team. The power engineer will undoubtedly acquire a sense of achievement and pride when he displays his license on the wall of his work area.

CLASSIFICATION OF LICENSES

There are five levels of licenses:
1. Chief Engineer
2. First Class Engineer
3. Second Class Engineer
4. Third Class Engineer
5. Fourth Class Engineer

REGULATIONS GOVERNING EXAMINATIONS

An applicant for examination for such licenses shall be required to fulfill the following general requirements for all classifications.

1. Complete the required form, stating accurately and in detail all the mechanical and electrical training and experience he has acquired relative to the examination for which he is a candidate.

2. Satisfy the Examining Board as to his character and ability.

3. Prove that he has attained the required age.

4. Take an affidavit as printed on application, attesting that his statements are true.

5. Answer examination questions presented and obtain a minimum percentage of correct responses.

6. Pay the required fee as prescribed by the Licensing Agency.

Examinations will be prepared with every attempt to insure fairness to the candidate. Accepted methods and theory of tests and measurement will be employed. Examinations will be kept valid by revision and up-dating.

CURRICULA FOR EXAMINATIONS

The board will establish a list of subjects covering the experiences and technical knowledge required for each classification. Examinations are based on the proper curricula for all grades.

By using these guides and test books suggested, any candidate can prepare for current examinations.

SOME QUESTIONS ANSWERED

1. **Question:** What is the basic difference between NIULPE and traditional licensing procedures?

   Answer: NIULPE attempts to establish a national standard for the licensing of power engineers. Existing licensing procedures establish requirements on a regional basis only, and, often, do not recognize other licensing agencies.

2. **Question**: Why is there a need for a system such as NIULPE?
   Answer: We are a mobile, migrant society when one considers the movement of personnel, the relocation of plants, and the branching out of industry, often in the form of subsidiaries. The need is based on a recognized attestation to a man's level of competency rather than on the local *modus operandi.*

3. **Question**: But will NIULPE go beyond state boundaries?
   Answer: The program is available nationwide on request. It follows, then, that if various states come under the program and examine and issue licenses as such, NIULPE's standard will be norm. You now have reciprocity.

4. **Question**: If this is so, how do you reconcile the great variance in certain aspects of power engineering, such as different fuels; the different specialities of various areas, such as refrigeration in the South, coal in one area, gas in another? There are many more.
   Answer: Anyone practicing the profession of power engineering will gain through research, study, conversation, and so on, a broad knowledge of the overall field. Obviously, his expertise will be greater within the areas of his immediate experience—fuels, heating, refrigeration, etc. However, the program is broad enough to cover these various areas and is based on a man's ability to manage energy conversion wherever he may practice.

5. **Question**: The same holds true when comparing existing license classifications. For instance, there may be five grades of license in one area, only two or three in another. NIULPE calls for five. How do you expect to gain entry and acceptance of this national uniformity of grades where there is now a variance?
   Answer: This is a classic problem of standardization that has persisted throughout the history of standards. Ideally, each plant should have a licensing program particular to its own operation and conditions. Obviously, this is impractical and does not lend itself to formulation of a standard. After researching the various licensing programs throughout this country and Canada, it is believed that

operating personnel can best be classed in one of five grades of license, namely, those we project in NIULPE: chief engineer, first class engineer, second class engineer, third class engineer, fireman and watertender. Actually, there is no conflict between NIULPE grading and that in those areas now having fewer grades of experience and education into a stated level of competency.

6. **Question**: What do you anticipate will be the reception of NIULPE in areas where codes now exist? For instance, at least one state has had a state-wide code since the turn of the century.

   Answer: In these areas the NIULPE program may at first be viewed as a form of competition. Existing agencies may fear an attempt to usurp their prerogatives. In reality, nothing is further from the truth. NIULPE is, in fact, an assisting agency to licensing agencies. It does not issue licenses. The program is logical and practical. It can supplement and upgrade if and as needed. It is conceivable that some programs now operating in various areas may parallel ours. Here there will be little if any need for modification of the program to come within NIULPE's national standards. There will be changes, more in some areas, less in others. Here is where NIULPE can be of great assistance.

7. **Question**: What about those now holding licenses?

   Answer: An existing licensing agency seeking to transfer its licenses to the five aforestated classes of license would determine, as its prerogative, the class of license that each licensee would receive, being guided by the standards of NIULPE.

8. **Question**: Is NIULPE for and by the National Association of Power Engineers? For members only, so to speak?

   Answer: Yes and no. Coincidentally, its organizers were and remain members of NAPE, and it is hoped that most, if not all, of our members will seek licenses under agencies certified in the NIULPE program. However, the program is open to any agency licensing power engineers. This, then, means the program is available to anyone in the profession.

9. **Question**: What is the relationship between NAPE and NIULPE?

   Answer: As aforementioned, NIULPE was conceived and organized by engineers who were and are now members of NAPE. Drawing on NAPE's 90-year history, it is logical to assume and is our hope that NAPE members will continue to be active and interested in NIULPE. Further, because of their geographical spread and collective knowledge of power engineering, NAPE members will continuously play a leading role in maintaining and upgrading NIULPE.

10. **Question**: What does NAPE expect to gain from this program?

    Answer: Materially, probably nothing. It will be recognized, however, that any program of benefit to the profession of power engineering will axiomatically benefit NAPE and its members.

11. **Question**: You have stated that NIULPE is an assisting agency and does not license. How does it assist licensing agencies?

    Answer: First, as founders of the program as well as its researchers and developers, those now in NIULPE are the most knowledgeable in its concept and operation, at least at this time. Second, NIULPE is in the key position to suggest and advise. Third, as mentioned earlier, NIULPE offers its knowledge and resources to the extension of its program through the local agency.

12. **Question**: How does NIULPE control the program?

    Answer: Any control exerted by NIULPE is on the local licensing agency thusly:
    - a. It promotes a standard for power engineers.
    - b. It establishes a qualification for examining engineers.
    - c. It establishes qualifications for instructors in power technology.
    - d. It accredits courses in power technology predicated on the education necessary for license examination.
    - e. It establishes uniform examinations for the various classes of license and offers those uniform examinations to licensing agencies.

13. **Question**: Do you anticipate NIULPE in its present form, or with some variance, ever becoming state or local law?

Answer: A complex question. We do not foresee NIULPE becoming law in any state. We do, however, foresee states creating legislation to form license laws under the NIULPE program. It must be remembered that NIULPE is an engineering standard and, as such, leaves the legislative function to agencies created by that legislation.

14. Question: Is the position just described desirable?

Answer: Yes. A law is difficult to change. Very few professions advance in technology as rapidly as engineering. With the energy crisis and our environmental and ecological problems, we may expect this technology to advance at an even more rapid pace. A law written for today's technology may very well be totally obsolete within months. If NIULPE is to promote realistic standards for the profession of power engineering, it must be flexible enough to remain current with ever advancing technology.

15. Question: What is the structure of NIULPE; what personnel are involved?

Answer: NIULPE is a non-profit corporation consisting of the usual officers and board of directors required by law. Membership in NIULPE is limited to the chief engineering examiner of every licensing agency operating under the rules of NIULPE. This chief examining engineer within his licensing agency at the local level has a team of examining engineers, one of whom shall be secretary of this local board of examining engineers. The chief examiner and his examining engineers handle all licensing actions at the level of their jurisdiction.

16. Question: Presently, the program is enjoined primarily by NAPE members. What is the plan to extend beyond NAPE membership? For examining engineers as well as license applicants?

Answer: It is logical to assume that the majority of those presently participating are members of NAPE. However, as the program becomes widely known and accepted by licensing agencies, those existing and those to be formed, it is only natural that members of the profession of power engineering, no matter their affiliation, will be attracted and become participants through their normal aspirations to exhibit their respective level of competency through achievement of a higher class of license.

17. **Question:** How does an applicant for license become a part of the program?

    Answer: This is the function of the local licensing agency. Application for license is voluntary on the part of the applicant. But it is to be hoped that licensing agencies will be most diligent in seeking to offer this attestment of an established level of competency to all personnel in their respective areas who are practicing the profession of power engineering.

18. **Question:** How does an applicant for a license know what steps he must take to prepare himself for the grade of license he seeks?

    Answer: Generally speaking, an applicant does not seek a specific class of license. It is the duty of the licensing agency to suggest an appropriate course of action for determining the applicant's experience and educational background. Each licensing agency should make its program known to all personnel in the profession of power engineering within its jurisdiction.

19. **Question:** What about the "grandfather clause?"

    Answer: NIULPE takes no position in this matter. Realizing the varying conditions from region to region, NIULPE leaves this to each NIULPE state board of examining engineers. Each board should establish ground rules under which licenses will be issued. Once a license has been issued, any advancement in grade can be achieved only through written examination. Reciprocity is granted between regions on certificates issued after successful examination.

20. **Question:** What about jurisdiction: For instance, an applicant from one state, which does not have the NIULPE program, desires to enter the program?

    Answer: He may be examined by the Board of another state, and if successful in qualifying, will be issued a license in the appropriate grade in the name of the issuing board.

21. **Question**: What is the calendar of expectations for this program? How widely will it be accepted?

Answer: As the singular purpose of this program is to recognize and attest to an established level of competency, as a man's competence and the job safety that can be expected from his endeavors are in direct relation, and with the impetus nationally on job safety, it follows that the program will be in operation, or at least recognized, in any area of the United States where job safety is deemed important.

22. **Question**: How does NIULPE enter into training and instruction and curriculum thereto?

Answer: This text contains the subject areas involved in successful qualification for each of the five grades of licensing. Study by the individual and group instruction should be directed to these subject areas. Through the commissioning of instructors and accreditation of schools, NIULPE points toward standardization of courses, texts and instruction. It is cooperating with special committees of NAPE in the assembly of suitable texts and dissemination of suggested curriculum.

23. **Question**. How does NIULPE enter in to the training of apprentices?

Answer: Apprenticeship is not part of the NIULPE function. However, recognizing that the average age of the power engineer today is 55, this means that in 10 years on the average, 100% of our power engineers must be replaced. Apprenticeship training is one of the vehicles available for training future power engineers. Since these future power engineers may desire to attest to their level of competency through a license, it clearly follows that the professional standards promoted by NIULPE can realistically be expected to form the outline for an apprenticeship training program.

SUMMARY OF INTENT

The National Institute for the Uniform Licensing of Power Engineers will:

1. Establish a voluntary standardization and certification program on a national scale working through and with existing licensing bodies.
2. Establish such programs where none now exists.
3. Establish five grades of licenses.
4. Allow for reciprocity between states.
5. Assist states to write licensing programs.
6. Disseminate information and standards.
7. Revise and update as technology demands.
8. Commission examining engineers.
9. Commission instructors.
10. Evaluate and issue credentials of accreditation to those instructional agencies meeting criteria established.
11. Permit states not now having state codes to obtain same.
12. Place into effect standards and guidelines authored by those within the profession of power engineering.
13. Serve all in power engineering regardless of affiliation.

APPENDIX B
Associated Organization Summary

ASSOCIATED ORGANIZATION SUMMARY

A.B.M.A.	A.S.M.E.	N.A.P.E.
A.E.E.	A.S.T.M.	N.B.B.P.V.I.
A.G.A.	B.O.C.A.	N.E.M.A.
A.I.S.E.	B.O.M.A.	N.F.P.A.
A.N.S.I.	B.O.M.I.	O.S.H.A.
A.P.I.	I.B.R.	S.A.E.
A.S.H.R.A.E.	I.U.O.E.	U.L.

A.B.M.A.

American Boiler Manufacturers Association
950 N. Glebe Road, Suite 160
Arlington, VA 22203

An association comprised of boiler manufacturers which sponsors training programs and compiles statistics on boiler sales and fuel consumption.
(membership 97)

A.E.E.

Association of Energy Engineers
4025 Pleasantdale Road, Suite 420
Atlanta, GA 30340

An association of engineers and persons interested in the cost-effective application and conservation of energy which promotes awareness of the topic through education and judicious use of our natural resources.

A.G.A.

American Gas Association
1515 Wilson Boulevard,
Arlington, VA 22209

A voluntary trade association consisting of representatives from utilities, manufacturers and other organizations within the gas industry which sets standards for the protection of consumers. (membership 4,758)

A.I.S.E.

Association of Iron and Steel Engineers
3 Gateway Center, Suite 2350
Pittsburgh, PA 15222

Association comprised of engineers, operators and suppliers organized to advance the production and processing of iron and steel who conduct studies and supply technical reports for the industry.
(membership 12,000)

A.N.S.I.

American National Standards Institute, Inc.
1430 Broadway
New York, NY 10018

An independent organization that identifies industrial and public requirements for national consensus standards and coordinates and manages their development, resolves national standards problems, and ensures effective participation in international standardization. (membership 1,100)

A.P.I.

American Petroleum Institute
1220 L Street, N.W.
Washington, DC 20005

An organization consisting of producers, refiners and transporters of petroleum-based products and natural gas that encourages the study of arts and sciences connected with the petroleum industry. (membership 6,440)

A.S.H.R.A.E.
American Society of Heating, Refrigeration
and Air Conditioning Engineers
1791 Tullie Circle, N.E.
Atlanta, GA 30329

A professional society comprised of HVAC engineers which sponsors research programs in co-operation with universities, laboratories and government agencies on the effects of air-conditioning, heat transfer, quality of air . . . etc.
(membership 50,000)

A.S.M.E.
American Society of Mechanical Engineers
345 E. 47th Street
New York, NY 10017

An educational association dedicated to the advancement of mechanical engineering which organizes and supports committees that set the standards for the practice. The A.S.M.E. does not approve, certify or endorse any product or construction.
(membership 103,918)

A.S.T.M.
American Society for Testing and Materials
1916 Race Street
Philadelphia, PA 19103

An organization which develops standards for the specification, testing and definition of materials, then adopts them based on the consensus vote of its membership which is comprised of industry committees, representatives of special interest groups, consumers, manufacturers and members of the scientific community.
(membership 30,050)

B.O.C.A.
Building Officials and
Code Administrators International, Inc.
4051 W. Flossmoor Road
Country Club Hills, IL 60477

An organization comprised of government officials and agencies that promotes establishment of unbiased building codes which supplies information on the acceptability of building materials, conducts seminars and assists in preparing in-service training programs for local organizations.
(membership 7,000)

B.O.M.A.

Building Owners and Managers Association
1250 Eye Street, N.W., Suite 200
Washington, DC 20005

An association consisting of owners, managers and developers of commercial office buildings that promotes dissemination of relevant information and establishes standards of performance for the industry.
(membership 5,400)

B.O.M.I.

Building Owners and Managers
Institute International, Inc.
P.O. Box 9709
1521 Ritchie Highway, Suite 3A
Arnold, MD 21012

The educational arm of B.O.M.A. providing programs toward certification as Systems Maintenance Technician and Systems Maintenance Administrator in the power plant field.

I.B.R.

Institute of Boiler and Radiator Manufacturers
(Hydronics Institute)
35 Russo Place
Berkeley Heights, NJ 07922

An institute comprised of manufacturers of heating and cooling equipment which sponsors educational programs in selected cities.
(membership 75)

I.U.O.E.

International Union of Operating Engineers
1125 17th Street N.W.
Washington, DC 20036

An organization consisting of individuals working in the plant engineering field which promotes training and advancement of its members throughout the industry.
(membership 420,000)

N.A.P.E.

National Association of Power Engineers, Inc.
2350 E. Devon Avenue
Des Plaines, IL 60018

A professional society comprised of Stationary Engineers which promotes education in the power engineering field and sets standards for licensing of operators.
(membership 10,000)

N.B.B.P.V.I.

National Board of Boiler and Pressure Vessel Inspectors
1055 Crupper Avenue
Columbus, OH 43229

A voluntary organization made up of chief inspectors from political subdivisions of the United States and Canada that promotes the uniform enforcement of boiler and pressure vessel laws and rules.
(membership 50)

N.E.M.A.

National Electrical Manufacturers' Association
155 East 44th Street
New York, NY 10017

An organization of electrical manufacturers that establishes standards for manufacture and tests for performance and reliability of electrical products. N.E.M.A. tests are often the basis or prerequisite for approval by the Federal government and/or by Underwriters' Laboratories, Inc.

N.F.P.A.
National Fire Protection Association
470 Atlantic Avenue
Boston, MA 02210

An organization devoted to promoting the science and improving the methods of Fire Protection. Membership is open to anyone interested. Every 3 years the N.F.P.A. produces a new edition of the NEC. The N.F.P.A. also provides many other useful publications dealing with fire prevention.

O.S.H.A.
Occupational Safety and Health Administration
U.S. Department of Labor
200 Constitution Ave. N.W.
Washington DC 20210 (puls regional offices)

That part of the U.S. Department of Labor responsible for assuring that employers provide safe and healthful working conditions and equipment for employees, and that employees properly avail themselves of these conditions. O.S.H.A. does not approve products. Compliance with O.S.H.A. regulations is contingent on approval or listing of the product by an authorized testing laboratory such as UL and proper installation and/or use of the product in accordance with O.S.H.A. guidelines.

S.A.E.
Society of Automotive Engineers
400 Commonwealth Drive
Warrendale, PA 15096

A society formed to advance the arts, sciences, standards and engineering practices related to the design, construction and utilization of self-propelled mechanisms, prime movers and their components.
(membership 43,000)

UL
Underwriters Laboratories, Inc.
Chicago, Northbrook, IL; Melville, NY; Santa Clara, CA

An independent, not-for-profit organization testing for public safety. Tests by UL are bases for acceptance by various government agencies. Listing by UL denotes initial testing and often periodic retesting for safety of operation.

APPENDIX C

Chemicals Used In Water Treatment

(Courtesy: Betz Laboratories)

NAMES Common, trade and chemical CHEMICAL FORMULA FORMULA WEIGHT	GRADES Usually used in water treatment	SOLUBILITIES in parts per 100 parts of water 32°F 50°F 68°F 86°F	SHIPPING CONTAINER	WEIGHT lb. per cu. ft.	STORAGE SPACE cu. ft. per ton
ACIDS					
HYDROCHLORIC ACID MURIATIC ACID *Formula:* HCl *Formula weight:* 36.46	18° Baume = 27.92% HCl 20° Baume = 31.45% HCl 22° Baume = 36.00% HCl	Miscible with water in all proportions	Bottles Carboys Drums Tank trucks Tank cars	1 U.S. gal. 18° Be weighs 9.53 lb. 1 cu. ft. 18° Be weighs 71.1 lb. 1 U.S. gal. 20° Be weighs 9.65 lb. 1 cu. ft. 20° Be weighs 72.3 lb. 1 U.S. gal. 22° Be weighs 9.83 lb. 1 cu. ft. 22° Be weighs 73.5 lb.	
PHOSPHORIC ACID ORTHO-PHOSPHORIC ACID *Formula:* H_3PO_4 *Formula weight:* 98.0	75% H_3PO_4 Other grades are available in syrup or paste form	Miscible with water in all proportions	Carboys Kegs Barrels (paste) Tank cars	1 U.S. gal. 75% H_3PO_4 weighs 13.1 lb. 1 cu. ft. 75% H_3PO_4 weighs 98.2 lb.	
SULFURIC ACID OIL OF VITRIOL *Formula:* H_2SO_4 *Formula weight:* 98.1	60° Baume = 77.67% H_2SO_4 66° Baume = 93.19% H_2SO_4	Miscible with water in all proportions	Bottles Carboys Drums Tank trucks Tank cars	1 U.S. gal. 60° Be weighs 14.2 bl. 1 cut. ft. 60° Be weighs 106.3 lb. 1 U.S. gal. 66° Be weighs 15.3 lb. 1 cu. ft. 66° Be weighs 114.4 lb.	

ALKALIES

BICARBONATE OF SODA BAKING SODA SODIUM BICARBONATE *Formula:* NaHCO$_3$ *Formula weight:* 84.0	Powder: 99% NaHCO$_3$	6.5 7.5 8.8 10.0	Bags Barrels Kegs In bulk	59 to 62	32 to 34
CAUSTIC SODA LYE SODA LYE SODIUM HYDROXIDE *Formula:* NaOH *Formula weight:* 40.0	Flake: 76% Na$_2$O = 98.06% NaOH Solid: 76% Na$_2$O = 98.06% NaOH Also available in ground form and as 50% to 73% liquid in tank cars	42.0 51.5 109 119	Flake: Drums Solid: Drums	400 lb. drum 100 lb. drum 50 lb. drum 700 lb. to 730 lb. drums	21.25" dia. x 32.25" 15.6" dia. x 17.25" 12.5" dia. x 13.5" 21.25" dia. x 32.25" to 33.75" high
SILICATE OF SODA WATER GLASS SODIUM SILICATE *Formula corresponds approximately to* Na$_2$O 3.25 SiO$_2$	Grade usually used is 40° Baume solution Solid glass and grades with varying ratios of Na$_2$O to SiO$_2$ also available	40° Be solution is miscible with water in all proportions	Drums Tank trucks Tank cars	1 U.S. gal 40° Be solution weighs 11.5 lb 1 cu. ft. 40° Be solution weighs 86.1 lb.	
SAL SODA WASHING SODA CRYSTAL SODIUM CARBONATE *Formula:* Na$_2$CO$_3$ · 10 H$_2$O *Formula weight:* 286	Crystal or lump required for pot type chemical feeds Soda Ash may be used instead in other types of feeds	18.9 33.8 58.0 105	Bags Kegs Barrels	68 to 71	28 to 30
SODA ASH SODIUM CARBONATE *Formula:* Na$_2$CO$_3$ *Formula weight:* 106	Dense soda ash: Powder Light soda ash: Powder For both dense & light 58% Na$_2$O = 99.16% Na$_2$CO$_3$	7.0 12.5 21.5 38.8	Bags Barrels In bulk	Dense 68 to 78 Light 35 to 46 The 100 lb. paper bag occupies, when stacked, about 1.45 cu. ft. for dense soda ash and 2.36 cu. ft. for light soda ash.	29 to 53

NAMES — Common, trade and chemical / CHEMICAL FORMULA / FORMULA WEIGHT	GRADES — Usually used in water treatment	SOLUBILITIES in parts per 100 parts of water 32°F 50°F 68°F 86°F	SHIPPING CONTAINER	WEIGHT lb. per cu. ft.	STORAGE SPACE cu. ft. per ton
BARIUM CARBONATE					
WITHERITE / Formula: $BaCO_3$ / Formula weight: 197	Powdered	0.002 0.002 0.002 / Due to low solubility is usually fed dry or in suspension	Bags / Barrels / In bulk	52 to 78	26 to 39
CALCIUM CARBONATE, CHLORIDE and SULPHATE					
CALCIUM CARBONATE / WHITING / CHALK / LIMESTONE / CALCITE / Formula: $CaCO_3$ / Formula weight: 100	Crushed and graded granules are used as a filter medium / Powder is seldom employed in water treatment	0.0014 in pure CO_2 – free water at 60°F.	Bags / Barrels / In bulk	*Powder* / 48 to 71 / *Fine Granules* / 100 to 115	28 to 42 / 18 to 20
CALCIUM CHLORIDE / Flake: 77%–78% $CaCl_2$ / Solid: 73%–75% $CaCl_2$ / Liquid: 40% $CaCl_2$ / Formula: $CaCl_2$ (anhydrous) / Formula weight: 111 (anhydrous)		59.5 65.0 74.5 102	Flake: Moisture proof bags or drums / Solid: Drums / Liquid: Tank cars	*Flake* / 66 to 69 / 400 lb. drums / *Solid* / 700 lb. drums / Liquid—1 U.S. gal. weights 11.6 lb. / Liquid—1 cu. ft. weighs 87.0 lb.	*Flake* / 29 to 31 / 24¼" dia. x 33¼" / *Solid* / 21¼" dia. x 33¾"
GYPSUM / CALCIUM SULFATE / Formula: $CaSO_4 . 2H_2O$ / Formula weight: 172	Ground	0.18 0.19 0.20 0.21 / Due to limited solubility, is usually fed dry or in suspension	Bags / Barrels / In bulk	50 to 60	33 to 40
CARBON AND CLAY					
CARBON, ACTIVATED / Various trade names	Powder for mixing with water / Granular for filter medium	Insoluble / Usually fed dry or in suspension	Paper bags	*Powder* / 15 to 28 / *Granular* / 12 to 24	72 to 134 / 84 to 167
CLAY / KAOLIN	Ground, powdered, air-floated / Various degrees of fineness / Too fine a grade is often undesirable	Insoluble / Usually fed dry or in suspension	Bags / Barrels / In bulk	30 to 69	29 to 67

COAGULANTS

Name / Formula	Description	Solubility	Packaging	Weight	Slab, lump or ground (angle)
ALUMINA SULFATE / ALUMINUM SULFATE / FILTER ALUM Formula: $AL_2(SO_4)_3 \cdot 18\,H_2O$ Formula weight: 666	Slab, lump, ground or powdered. Iron-free grade not required. Acid to basic grades contain 14.5% to 17.5% Al_2O_3	60.8 65.3 71.0 78.8	Bags Kegs Barrels In bulk	57 to 67 38 to 45	30 to 35 (Powdered) 45 to 53
AMMONIA ALUM / AMMONIUM ALUM Formula: $Al_2(SO_4)_3 \cdot (NH_4)_2SO_4 \cdot 24\,H_2O$ Formula weight: 906	Lump or crystal required for pot type chemical feeds	3.9 9.5 15.1 20.0	Bags Kegs Barrels Boxes Carton	64 to 68	30 to 31
POTASH ALUM / POTASSIUM ALUM Formula: $Al_2(SO_4)_3 \cdot K_2SO_4 \cdot 24\,H_2O$ Formula weight: 949	Lump or crystal required for pot type chemical feeds	5.7 7.6 11.4 16.6	Bags Kegs Barrels Boxes Cartons	64 to 68	30 to 31
FERRIC SULFATE / FERRISUL / FERRIFLOC Formula: $Fe_2(SO_4)_3$ Formula weight: 400	Granules. Composition from different sources varies (70% to 90% $Fe_2(SO_4)_3$. Certain grades also contain $Al_2(SO_4)_3$	Very soluble. If cold water is used to dissolve, use 2 parts water to 1 part of ferric sulfate	Bags Kegs Barrels Drums In bulk	60 to 70	29 to 34
FERROUS SULFATE / COPPERAS Formula: $FeSO_4 \cdot 7H_2O$ Formula weight: 278	Crystals Granules Sugar sulfate	28.7 37.5 48.5 60.2	Bags Boxes Kegs Barrels In bulk	63 to 66	30 to 32
SODIUM ALUMINATE Formula: $Na_2Al_2O_4$ Formula weight: 163.9	Crystal or granules: 67-90% $Na_2Al_2O_4$, 4% excess caustic. Liquid: 32% $Na_2Al_2O_4$, 7% caustic	29.5 34.0 37.0 40.0	Bags Drums Barrels	50 to 60	33 to 40

COPPER SULPHATE

Name / Formula	Description	Solubility	Packaging	Weight	Slab, lump or ground (angle)
COPPER SULFATE / BLUESTONE / BLUE VITRIOL Formula: $CuSO_4 \cdot 5H_2O$ Formula weight: 250	Crystals Granules Powder	19.5 23.1 26.9 31.3	Bags Barrels	75 to 87	23 to 27

NAMES Common, trade and chemical CHEMICAL FORMULA FORMULA WEIGHT	GRADES Usually used in water treatment	SOLUBILITIES in parts per 100 parts of water				SHIPPING CONTAINER	WEIGHT lb. per cu. ft.	STORAGE SPACE cu. ft. per ton
		32°F	50°F	68°F	86°F			
GASES								
AMMONIA Formula: NH₃ Formula weight: 17.0	Liquid under pressure Ammonium hydroxide solution (NH₄OH)—°Be, 29.4% NH₃ Aqua Ammonia	47.3	40.6	34.6	29.1 Above solubilities at atmospheric pressure	Steel cylinders 50 lb. net 100 lb. net Tank cars 50,000 lb. net	Gross weight 117 lb. 233 lb.	8¼" dia. x 53" 12½" dia. x 55"
CHLORINE Formula: Cl₂ Formula weight: 70.9	Liquid under pressure	1.46	0.980	0.716	0.562 Above solubilites at atmospheric pressure	Steel cylinders 105 lb. net 150 lb. net 2000 lb. net Tank cars Multi-unit Single-unit	Gross weight 195 lb. 270 lb. 3325 lb. Net weight Fifteen 1 ton units— 32,000 to 60,000 lb.	10½" dia. x 40" 10½" dia. x 52½" 30" dia. x 80"
SULPHUR DIOXIDE Formula: SO₂ Formula weight: 64.1	Liquid under pressure	22.8	16.2	11.3	7.8	Steel cylinders 100 lb. net 150 lb. net 2000 lb. net Tank cars Multi-unit Single-unit	Gross weight 177 lb. 221 lb. 3369 lb. Net weight Fifteen 1 ton units 40,000 lb.	8¼" dia. x 53" 10" dia. x 53" 30" dia. x 80"

LIMES AND MAGNESIA

Material	Description	Slaking / Solubility	Packaging	Weight	Range
CHEMICAL LIME / QUICK LIME / BURNT LIME / CALCIUM OXIDE — *Formula:* CaO — *Formula weight:* 56.1	Lump, pebble or ground High calcium lime usually contains 90% CaO	Slakes with water forming hydrated lime, the solubility of which is as follows: 0.18 0.17 0.16 0.15 To slake, use about ½ gal water per pound of quick lime	Bags Barrels In bulk	Lump 50 to 65 Pebble 60 to 65 Ground 50 to 70 Pulverized 39 to 71	31 to 40 31 to 34 29 to 40 28 to 52
HYDRATED LIME / SLAKED LIME / CALCIUM HYDROXIDE — *Formula:* Ca(OH)$_2$ — *Formula weight:* 74.1	Powder Usually contains 93% Ca(OH)$_2$	0.18 0.17 0.16 0.15	Bags Barrels In bulk	25 to 50	40 to 80
DOLOMITIC LIME — *Formula:* CaO + MgO Content of MgO varies	Lump, pebble or ground Typical analysis for silica removal—58% CaO and 40% MgO	Slakes with water forming hydrated lime and magnesium oxide the latter of which slakes very slowly	Bags Barrels In bulk	Lump 50 to 65 Pebble 60 to 65 Ground 50 to 75 Pulverized 37 to 63	31 to 40 31 to 34 27 to 40 32 to 54
HYDRATED DOLOMITIC LIME — *Formula:* Ca(OH)$_2$ + MgO Content of MgO varies	Powder Typical analysis for silica removal—62% Ca(OH)$_2$ and 32% MgO		Bags Barrels In bulk	28 to 52	39 to 72
MAGNESIA / MAGNESIUM OXIDE — *Formula:* MgO — *Formula weight:* 40.3	Powder Various grades differ greatly in density	Hydrates very slowly in water Solubility of hydrate very low as shown below 0.002 0.002	Bags Barrels In bulk	8 to 40	50 to 250

PHOSPHATES OF SODA

Material	Solubility	Packaging	Weight	Range
SODIUM PHOSPHATE MONOBASIC — *Formula:* NaH$_2$PO$_4$ (anhydrous) — *Formula weight:* 120 (anhydrous)	57.9 69.9 85.2 107.0		55 to 70	29 to 37
Formula: NaH$_2$PO$_4$. H$_2$O (monohydrate) — *Formula weight:* 138 (monohydrate)	66.6 80.4 98.0 123.0	Bags Barrels Kegs Boxes Cans Drums	58 to 79	26 to 35
DIBASIC — *Formula:* Na$_2$HPO$_4$ (anhydrous) — *Formula weight:* 142 (anhydrous)	1.7 3.6 7.7 20.8		53 to 62	32 to 39
Formula: Na$_2$HPO$_4$. 12H$_2$O (crystalline) — *Formula weight:* 358 (crystalline)	4.2 8.9 19.3 52.4		46 to 53	38 to 44

NAMES Common, trade and chemical CHEMICAL FORMULA FORMULA WEIGHT	GRADES *Usually used in water treatment*	SOLUBILITIES in parts per 100 parts of water 32°F 50°F 68°F 86°F	SHIPPING CONTAINER	WEIGHT lb. per cu. ft.	STORAGE SPACE cu. ft. per ton
PHOSPHATES OF SODA (continued)					
TRIBASIC *Formula:* $Na_3PO_4 \cdot H_2O$ (monohydrate) *Formula weight:* 182 (monohydrate)		1.7 4.6 12.3 22.2		83 to 90	22 to 24
Formula: $Na_3PO_4 \cdot 12H_2O$ (crystalline) *Formula weight:* 380 (crystalline)		3.5 9.5 25.5 46.4		56 to 60	32 to 36
META (Hexa) *Formula:* $NaPO_3$ *Formula weight:* 102		Very soluble 25% to 50% solutions very viscous Usually fed in solutions of 1 to 2 lb. per U.S. gal.		17 to 19 *Flake* 39 to 41 *Glass*	105 to 118 49 to 51
SODIUM CHLORIDE, SULFATES AND SULFITE					
SALT COMMON SALT SODIUM CHLORIDE *Formula:* NaCl *Formula weight:* 58.5	Rock Salt Evaporated Salt 98% NaCl	35.7 35.8 36.0 36.3	Bags Barrels In bulk	50 to 70 Evaporated near lower limits Rock near upper limits	29 to 40
SALT CAKE SODIUM SULFATE *Formula:* Na_2SO_4 (anhydrous) *Formula weight:* 142 (anhydrous)	Lumps Powder Usually 92% to 99% Na_2SO_4	5.0 9.0 19.4 40.8	Bags Barrels In bulk	85 to 95	21 to 24
SODIUM SULFATE CRYSTALLINE GLAUBER'S SALT *Formula:* $Na_2SO_4 \cdot 10H_2O$ *Formula weight:* 322	Crystals	11.3 20.4 44.0 92.5	Bags Barrels Boxes In bulk	58 to 63 *Standard Crystal* 46 to 56 *Needle Crystal*	32 to 35 36 to 44
SODIUM BISULFATE NITRE CAKE SODIUM ACID SULFATE *Formula:* $NaHSO_4$ *Formula weight:* 120	Lump Ground	Very soluble At 68°F 100 parts of water will dissolve about 30 parts of $NaHSO_4$	Barrels In bulk	84 to 89 *Lump* 85 to 95 *Ground*	23 to 24 21 to 24
SODIUM SULFITE *Formula:* Na_2SO_3 *Formula weight:* 126	Crystals Powder	14.0 20.0 27.0 36.0	Bags Kegs Barrels Drums	80 to 91	23 to 26

Sodium Chloride Solutions

Calculated from Gerlach's tables

Percent NaCl or grams per 100 grams of solution	sp. gr. at 15° C (15° C or 59° F) / 59° F	° Baume at 15° C or 59° F	lb. NaCl per U.S. gal.	lb. NaCl per cu. ft.
1.0	1.0073	1.0	0.084	0.63
2.0	1.0145	2.0	0.169	1.27
3.0	1.0217	3.1	0.255	1.91
4.0	1.0290	4.1	0.343	2.57
5.0	1.0362	5.1	0.432	3.23
6.0	1.0437	6.1	0.522	3.90
7.0	1.0511	7.1	0.612	4.59
8.0	1.0585	8.0	0.705	5.28
9.0	1.0659	9.0	0.799	5.98
10.0	1.0734	9.9	0.894	6.69
11.0	1.0810	10.9	0.990	7.41
12.0	1.0886	11.8	1.09	8.14
13.0	1.0962	12.7	1.19	8.88
14.0	1.1038	13.6	1.29	9.63

Percent NaCl or grams per 100 grams of solution	sp. gr. at 15° C (15° C or 59° F) / 59° F	° Baume at 15° C or 59° F	lb. NaCl per U.S. gal.	lb. NaCl per cu. ft.
15.0	1.1115	14.5	1.39	10.4
16.0	1.1194	15.5	1.49	11.2
17.0	1.1273	16.4	1.60	11.
18.0	1.1352	17.3	1.70	12.7
19.0	1.1432	18.2	1.81	13.5
20.0	1.1511	19.0	1.92	14.4
21.0	1.1593	19.9	2.03	15.2
22.0	1.1676	20.8	2.14	16.0
23.0	1.1758	21.7	2.25	16.9
24.0	1.1840	22.5	2.37	17.7
25.0	1.1923	23.4	2.48	18.6
26.0	1.2010	24.3	2.60	19.5
26.4	1.2043	24.6	2.65	19.8

NOTE: The above table is based on solutions of chemically pure sodium chloride. In regenerating zeolites, the strong commercial salt solutions obtained from brine tanks, salt saturators or wet salt storage basins, are assumed to contain 2.48 lb. sodium chloride per U.S. gallon or approx. 18.5 lb. per cu. ft.

Chemical Resistance Chart

	66°Bé H₂SO₄	dil. H₂SO₄	NaHSO₄ Sol'n	NaHSO₃ Sol'n	conc. HCl (36°)	dil. HCl	conc. HNO₃	dil. HNO₃	HF 50%	NaOH 50%	Paraffinics	Ketones	Amines	Alcohols	Aromatics
Metals															
Tantalum	A	A	A	A	A	A	A	A	X	X	—	—	—	—	—
20 SS	A	A	A	A	X	X	A	A	B	B	—	—	—	—	—
Hastelloy B	A	B	B	C	A	B	X	X	B	A	—	—	—	—	—
Hastelloy C	A	B	A	A	B	A	B	A	A	B	—	—	—	—	—
316 SS	B	B	B	B	X	X	A	A	C	A	—	—	—	—	—
304 SS	C	C	C	C	X	X	A	A	X	A	—	—	—	—	—
Cr-Si Steel (Durichlor)	A	A	A	X	B	A	B	B	X	C	—	—	—	—	—
LC Steel	B	X	X	X	X	X	X	X	X	A	—	—	—	—	—
Brass or Bronze (low zinc)	X	C	C	C	X	X	X	X	X	C	—	—	—	—	—
Copper	C	B	B	B	X	C	X	X	B	B	—	—	—	—	—
Aluminum	C	C	C	B	X	X	C	X	X	X	—	—	—	—	—
Plastics															
Polyethylene	A	A	A	A	A	A	B	A	A	A	B	C	C	A	C
Polypropylene	A	A	A	A	A	A	A	A	C	A	A	B	B	A	C
PVC (I or II)	A	A	A	A	A	A	B	A	A	A	A	C	C	A	X
Fiberglas (Atlac 382)	X	A	A	A	A	A	B	A	X	B	B	X	X	A	X
Penton	C	A	A	A	A	A	B	A	A	A	A	C	B	A	B
Kynor	A	A	A	A	A	A	A	A	A	A	A	C	B	A	A
Epoxy	C	A	A	A	A	A	C	A	B	A	A	C	—	A	A
Nylon	X	X	X	X	X	X	C	A	X	A	A	A	A	A	A
Plast. PVC (Tygon)	C	A	A	A	B	A	X	C	B	C	A	X	X	A	X

Material																	
Saran	C	A	A	A	A	A	A	A	A	C	A	X	X	X	A	A	C
Teflon	A	A	A	A	A	A	A	A	A	A	A	A	A	A	A	A	A
Rubbers																	
Viton	A	A	A	A	A	A	A	A	B	A	B	X	X	B	A	B	B
Neoprene	X	C	A	A	X	C	X	C	B	B	B	X	B	A	B	C	C
Buna N	X	C	A	A	X	C	X	C	B	C	B	X	A	X	B	A	C
Ethylene Propylene	X	B	A	A	C	A	C	B	A	B	A	A	B	A	A	B	B
Hypalon	B	A	A	A	B	A	B	A	A	A	A	B	A	A	B	A	B
Butyl	X	C	A	A	C	B	C	B	B	B	X	B	B	A	A	A	X
Silicone	X	C	A	A	X	X	X	C	B	X	X	X	X	B	A	A	B

NOTES:

Room Temperature Assumed —

A Excellent—No significant effect
B Good—Acceptable with few limitations
C Fair—Satisfactory under limited conditions
X Unsatisfactory
— No data

APPENDIX D
Periodic Table Of The Elements
(Courtesy: Betz Laboratories)

KEY TO CHART

Atomic Number → 50
Symbol → Sn
Atomic Weight → 118.69 Tin
Oxidation States

Transition Elements — Group VIII

Ia	IIa	IIIb	IVb	Vb	VIb	VIIb	VIII			Ib	IIb	IIIa	IVa	Va	VIa	VIIa	0
1 H 1.0079 Hydrogen																	2 He 4.00260 Helium
3 Li 6.94 Lithium	4 Be 9.01218 Beryllium											5 B 10.81 Boron	6 C 12.011 Carbon	7 N 14.0067 Nitrogen	8 O 15.9994 Oxygen	9 F 18.99840 Fluorine	10 Ne 20.17 Neon
11 Na 22.98977 Sodium	12 Mg 24.305 Magnesium											13 Al 26.98154 Aluminum	14 Si 28.086 Silicon	15 P 30.97376 Phosphorus	16 S 32.06 Sulfur	17 Cl 35.453 Chlorine	18 Ar 39.948 Argon
19 K 39.09 Potassium	20 Ca 40.08 Calcium	21 Sc 44.9559 Scandium	22 Ti 47.90 Titanium	23 V 50.941 Vanadium	24 Cr 51.996 Chromium	25 Mn 54.9380 Manganese	26 Fe 55.847 Iron	27 Co 58.9332 Cobalt	28 Ni 58.71 Nickel	29 Cu 63.546 Copper	30 Zn Zinc	31 Ga Gallium	32 Ge Germanium	33 As 74.9216 Arsenic (gray)	34 Se 78.96 Selenium	35 Br 79.904 Bromine	36 Kr 83.80 Krypton
37 Rb 85.467 Rubidium	38 Sr 87.62 Strontium	39 Y 88.9059 Yttrium	40 Zr 91.22 Zirconium	41 Nb 92.9064 Niobium (Columbium)	42 Mo 95.94 Molybdenum	43 Tc 98.9062 Technetium	44 Ru 101.07 Ruthenium	45 Rh 102.9055 Rhodium	46 Pd 106.4 Palladium	47 Ag Silver	48 Cd Cadmium	49 In 114.82 Indium	50 Sn 118.69 Tin	51 Sb 121.75 Antimony	52 Te 127.60 Tellurium	53 I 126.9045 Iodine	54 Xe 131.30 Xenon
55 Cs 132.9054 Cesium	56 Ba 137.34 Barium	57* La 138.9055 Lanthanum	72 Hf 178.49 Hafnium	73 Ta 180.947 Tantalum	74 W 183.85 Tungsten	75 Re 186.2 Rhenium	76 Os 190.2 Osmium	77 Ir 192.22 Iridium	78 Pt 195.09 Platinum	79 Au Gold	80 Hg Mercury	81 Tl 204.37 Thallium	82 Pb 207.2 Lead	83 Bi 208.9804 Bismuth	84 Po Polonium	85 At (210) Astatine	86 Rn (222) Radon
87 Fr (223) Francium	88 Ra 226.0254 Radium	89** Ac (227) Actinium	104 (261)	105 (262)	106 (263)												

*Lanthanides	58 Ce 140.12 Cerium	59 Pr 140.9077 Praseodymium	60 Nd 144.24 Neodymium	61 Pm (145) Promethium	62 Sm 150.4 Samarium	63 Eu 151.96 Europium	64 Gd 157.25 Gadolinium	65 Tb 158.9254 Terbium	66 Dy 162.50 Dysprosium	67 Ho 164.9304 Holmium	68 Er 167.26 Erbium	69 Tm 168.9342 Thulium	70 Yb 173.04 Ytterbium	71 Lu 174.97 Lutetium
**Actinides	90 Th 232.0381 Thorium	91 Pa 231.0359 Protactinium	92 U 238.029 Uranium	93 Np 237.0482 Neptunium	94 Pu (244) Plutonium	95 Am (243) Americium	96 Cm (247) Curium	97 Bk (247) Berkelium	98 Cf (251) Californium	99 Es (254) Einsteinium	100 Fm (257) Fermium	101 Md (256) Mendelevium	102 No (259) Nobelium	103 Lr (260) Lawrencium

Numbers in parentheses are mass numbers of most stable isotope of that element

APPENDIX E
Tables

LIST OF TABLES

AIR

CONVERSIONS

ELECTRIC

FORMULAE

HEAT

MISCELLANEOUS
23. Metric Capacity
24. Metric Length
25. Metric Weight
26. Weights of Metals
27. Ultimate Bending Strength
28. Recommended Lighting Levels

STEAM
29. Steam Pressure Temperature Relationship
30. Steam Trap Selection

TEMPERATURE
31. Color Scale of Temperature
32. Centigrade/Fahrenheit Scale
33. Melting Points of Common Substances
34. Temperatures of Waste Heat Gases

WATER
35. Water Equivalents
36. Supply Line Sizes for Common Fixtures
37. Pressure of Water
38. Water Requirements of Common Fixtures
39. Blowdown Flow Rates — 3-Inch Pipe
40. Tank Capacities Per Foot of Depth
41. Efficiency Loss Due to Scale

AIR
Table 1. Composition of Air

Nitrogen	78%
Oxygen	21%
Argon	0.96%
Carbon Dioxide & other gasses	0.04%

Table 2. Atmospheric Pressure Per Square Inch

Barometer	Pressure
28.00	13.75
28.25	13.88
28.50	14.00
28.75	14.12
29.00	14.24
29.25	14.37
29.50	14.49
29.75	14.61
29.921	14.696
30.00	14.74
30.25	14.86
30.50	14.98
30.75	15.10
31.00	15.23

Table 3. Approximate Air Needs of Pneumatic Tools, CFM

Grinders, 6- and 8-in. diameter wheels. .50
  2- and 2½-in. diameter wheels 14-20

File and burr machines 18

Rotary sanders, 9-in. diameter pads. 55

Sand rammers and tampers:
  1 x 4-in. cylinder. 25
  1¼ x 5-in. cylinder 28
  1½ x 6-in. cylinder 39

Chipping hammers, 10-13 lb 28-30
  2-4 lb 12

Nut setters to 5/16 in., 8 lb. 20
  ½ to ¾ in., 18 lb. 30

Paint spray 2-20

Plug drills 40-50

Riveters, 3/32- to 1/8-in. rivets 12

Steel drills, rotary motors:
  To ¼ in., weighing 1¼-4 lb 18-20

¼ to 3/8 in., weighing 6-8 lb 20-40
½ to ¾ in., weighing 9-14 lb 70
7/8 to 1 in., weighing 25 lb 80
Wood borers to 1-in. diameter,
  weighing 14 lb 40

Table 4. Average Absolute Atmospheric Pressure

Altitude in feet reference to sea level	Inches of Mercury (in. Hg)	Pounds per sq. in. absolute (psia)
− 1,000	31.00	15.2
− 500	30.50	15.0
sea level 0	29.92	14.7
+ 500	29.39	14.4
+ 1,000	28.87	14.2
+ 1,500	28.33	13.9
+ 2,000	27.82	13.7
+ 3,000	26.81	13.2
+ 4,000	25.85	12.7
+ 5,000	24.90	12.2
+ 6,000	23.98	11.7
+ 7,000	23.10	11.3
+ 8,000	22.22	10.8
+ 9,000	21.39	10.5
+ 10,000	20.58	10.1

CONVERSIONS

Table 5. Temperature Conversions

$°F = 9/5°C + 32$

$°F = °R − 459.58$

$°K = °C + 273.16$

$°R = °F + 459.48$

$°C = 5/9 (°F − 32)$

$°K = 5/9°F$

Table 6. Length and Area

1 statute mile (mi)	= 5280 feet
	= 1.609 kilometers
1 foot (ft)	= 12 inches
	= 30.48 centimeters
1 inch (in)	= 25.40 millimeters
100 ft per min	= 0.508 meter per sec
1 square foot	= 144 sq inches
	= 0.0929 sq meter
1 square inch	= 6.45 sq centimeters
1 kilometer (km)	= 1000 meters
	= 0.621 statute mile
1 meter (m)	= 100 centimeters (cm)
	= 1000 millimeters (mm)
	= 1,094 yards
	= 3.281 feet
	= 39.37 inches
1 micron	= 0.001 millimeter
	= 0.000039 inch
1 meter per sec	= 196.9 ft per min

Table 7. Horsepower Equivalent

1 HP = 33,000 ft. lb. per min
1 HP = 550 ft. lb. per sec.
1 HP = 2,546 B.t.u. per hr.
1 HP = 42.4 B.t.u. per min.
1 HP = .71 B.t.u. per sec.
1 HP = 746 watts

Table 8. Approximate Metric Equivalents

1 Decimetre	= 4 inches
1 Metre	= 1.1 yards
1 Hectare	= 2½ acres
1 Litre	= 1.06 qt.
1 Kilogramme	= 2.2 lb.
1 Metric Ton	= 2,200 lb.

Table 9. Miscellaneous Measures

Angles or Arcs

60 seconds (″). . . . . = 1 minute
60 minutes (′) . . . . = 1 degree
90 degrees (°) . . . . = 1 rt. angle or quadrant
360 degrees . . . . = 1 circle

Avoirdupois Weight

437.5 GRAINS (gr.) . . . . = 1 ounce
16 ounces (7,000) grains . . . . = 1 pound
2,000 pounds . . . . = 1 short ton
2,240 pounds . . . . = 1 long ton

Cubic Measure

2.728 cubic inches (cu. in.). . = 1 cubic foot
27 cubic feet . . . . = 1 cubic yard

Square Measure

144 square inches (sq. in.). . = 1 square foot
9 square feet . . . . = 1 square yard

ELECTRIC

Table 10. Appliance Energy Requirements

Major Appliances	Annual kWh
Air-Conditioner (room) (Based on 1000 hours of operation per year. This figure will vary widely depending on geographic area and specific size of unit)	860
Clothes Dryer	993
Dishwasher (including energy used to heat water).	2,100
Dishwasher only	363
Freezer (16 cu. ft.)	1,190
Freezer – frostless (16.5 cu. ft.)	1,820

Appliance Energy Requirements (con't,)

Major Appliances	Annual kWh
Range with oven	700
with self-cleaning oven	730
Refrigerator (12 cu. ft.)	728
Refrigerator — frostless (12 cu. ft.)	1,217
Refrigerator/Freezer (12.5 cu. ft.)	1,500
Refrigerator/Freezer — frostless (17.5 cu. ft)	2,250
Washing Machine — automatic (including energy used to heat water)	2,500
Washing Machine — non-automatic (including energy to heat water)	2,497
washing machine only	76
Water Heater	4,811

Kitchen Appliances

Blender	15
Broiler	100
Carving Knife	8
Coffee Maker	140
Deep Fryer	83
Egg Cooker	14
Frying Pan	186
Hot Plate	90
Mixer	13
Oven, Microwave (only)	190
Roaster	205
Sandwich Grill	33
Toaster	39
Trash Compactor	50
Waffle Iron	22
Waste Disposer	30

Heating and Cooling

Air Cleaner	216
Electric Blanket	147

Appliance Energy Requirements (con't.)

Major Appliances	Annual kWh
Dehumidifier	377
Fan (attic)	281
Fan (circulating)	43
Fan (rollaway)	138
Fan (window)	170
Heater (portable	176
Humidifier	163

Laundry

Iron (hand)	144

Health and Beauty

Germicidal Lamp	141
Hair Dryer	14
Heat Lamp (infrared)	13
Shaver	1.8
Sun Lamp	16
Toothbrush	.5
Vibrator	2

Home Entertainment

Radio	86
Television	
Black and White	
Tube type	350
Solid State	120
Color	
Tube type	660
Solid State	440

Housewares

Clock	17
Floor Polisher	15
Sewing Machine	11
Vacuum Cleaner	46

Table 11. Alternating Current Calculations

To Calculate	Alternating Current	
	Three-Phase	Single-Phase
Amperes when horsepower is known	$\dfrac{\text{H.P. x 746}}{\text{1.73 x E x \%Eff x P.F.}}$	$\dfrac{\text{H.P. x 746}}{\text{E x \%Eff x P.F.}}$
Amperes when kilowatts are known	$\dfrac{\text{K.W. x 1000}}{\text{1.73 x E x P.F.}}$	$\dfrac{\text{K.W. x 1000}}{\text{E x P.F.}}$
Amperes when K.V.A. are known	$\dfrac{\text{K.V.A. x 1000}}{\text{1.73 x E}}$	$\dfrac{\text{K.V.A. x 1000}}{\text{E}}$
Kilowatts	$\dfrac{\text{I x E x 1.73 x P.F.}}{1000}$	$\dfrac{\text{I x E x P.F.}}{1000}$
K.V.A.	$\dfrac{\text{I x E x 1.73}}{1000}$	$\dfrac{\text{I x E}}{1000}$
Horsepower (output)	$\dfrac{\text{IxEx1.73x\%EffxP.F.}}{746}$	$\dfrac{\text{I x E x \%Eff x P.F.}}{746}$

E = Volts. K.W. = Kilowatts. P.F. = Power Factor. I = Amperes
%Eff. = Percent Efficiency. K.V.A. = Kilovolt amperes.
H.P. = Horsepower.

Table 12. Kilowatt Conversion Factors

Kilowatt Conversion Factors

1 kilowatt	
=	1.3415 horsepower
=	738 ft lb per sec
=	44,268 ft lb per min
=	2,656,100 ft lb per hr
=	56.9 Btu per min
=	3,413 Btu per hr

Table 13. Effects of Electrical Current on Humans

Current Values	Effect
1 ma	Causes no sensation
1 to 8 ma	Sensation of shock. Not painful
8 to 15 ma	Painful shock
15 to 20 ma	Cannot let go
20 to 50 ma	Severe muscular contractions
100 to 200 ma	Ventricular fibrillation
200 & over ma	Severe burns. Severe muscular contractions

Table 14. Equivalent Electrical Units

1 Kilowatt. . . . . . . = 1,000 Watts

1 Kilowatt. . . . . . . = 1.34 H.P.

1 Kilowatt. . . . . . . = 44,260 Foot-Pounds per minute

1 Kilowatt. . . . . . . = 56.89 B.T.U. per minute

1 H.P. . . . . = 746 Watts

1 H.P. . . . . = 33,000 Foot-Pounds per minute

1 H.P. . . . . = 42.41 B.T.U. per minute

1 B.T.U. . . . . = 778 Foot-Pounds

1 B.T.U. . . . . = 0.2930 Watt-Hour

1 Joule . . . . = 1 Watt-Second

Table 15. United States Power Characteristics

	Voltage	Amperes	Phase
Controls	20 to 120	5 to 15	Single
Small Equipment	120		
	208		
	240	10 to 40	Single
	277		
Large Equipment	208		
	240	30 to 400	Three
	480		

Table 16. Induction Motor Synchronous Speeds

Poles	@ 60 Hz
2	3,600
4	1,800
6	1,200
8	900
10	720
12	600

FORMULAE

Table 17. Geometric Formulas

Circumference of a circle	$C = \pi d$
Length of an arc	$L = \dfrac{n}{360} \times \pi d$
Area of a rectangle	$A = LW$
Area of a square	$A = s^2$
Area of a triangle	$A = \frac{1}{2}bh$
Area of a trapezoid	$A = \frac{1}{2}h\,(b + b')$
Area of a circle	$A = .8754d^2$, or $\frac{1}{4}\pi d^2$
Area of a sector	$S = \dfrac{n}{360} \times .7854d^2$
Area of an ellipse	$A = .7854ab$
Area of the surface of a rectangular solid	$S = 2LW + 2LH + 2WH$
Lateral area of a cylinder	$S = \pi dh$
Area of the surface of a sphere	$S = \pi d^2$
Volume of a rectangular solid	$V = LWH$
Volume of a cylinder	$V = .7854d^2 h$
Volume of a sphere	$V = .5236d^3$, or $1/6\pi d^3$
Volume of a cube	$V = e^3$

HEAT

Table 18. Heat Equivalents

1 Btu	= 252 calories
1 kilocalorie	= 1000 calories
1 Btu/lb.	= .55 kcal/kg
1 Btu/lb.	= 2.326 kj/kg
1 kcal/kg	= 1.8 Btu/lb
1 Btu/hr	= 0.2931 watts

Table 19. Heat Generated by Appliances

General lights and heating	3.4 Btu/hr/watt
2650 watt toaster	9100 Btu/hr
5000 watt toaster	19,000 Btu/hr
Hair Dryer	2000 Btu/hr
Motor less than 2 HP	3600 Btu/hr/HP
Motor over 3 HP	3000 Btu/hr/HP

Table 20. Heat Loss from Hot Water Piping

Pipe Size, Inches	Hot Water, 180°F	
	Bare	Insulated
½	65	22
¾	75	25
1	95	28
1¼	115	33
1½	130	36
2	160	42
2½	185	48
3	220	53
4	280	68

Table 21. Specific Heats of Common Substances

Aluminum	.2143
Brine	.9400
Coal	.314
Copper	.0951
Ice	.5040
Petroleum	.5110
Water	1.0000
Wood	.3270

Table 22. Heat Content of Common Fuels

Number 6 fuel oil	152,400 Btu per gallon
Number 2 fuel oil	139,600 Btu per gallon
Natural Gas	950 to 1150 Btu per cubic foot
Propane	91,500 Btu per gallon

MISCELLANEOUS

Table 23. Metric Capacity

Name		Capacity
Milliliter (ml.=)	=	.001
Centiliter (cl.)	=	.01
Liter (l)	=	1.
Decaliter (Dl.)	=	10
Hectoliter (Hl.)	=	100
Kiloliter (Kl.)	=	1,000
Myrialiter (Ml.)	=	10,000

Table 24. Metric Length

		Meters
Millimeter (mm.)	=	.001
Centimeter (cm.)	=	.01
Decimeter (dm.)	=	.1
Meter (m)	=	1
Decameter (Dm)	=	10
Hectometer (Hm.)	=	100
Kilometer (Km.)	=	1,000
Myriameter (Mm.)	=	10,000

Table 25. Metric Weight

Name	Grams
Milligram (mg.)	.001
Centigram (cg.)	.01
Decigram (dg.)	.1
Gram (g)	1
Decagram (Dg.)	10
Hectogram (Hg.)	100
Kilogram (Kg.)	1,000
Myriagram (Mg.)	10,000
Quintal (Q.)	100,000
Tonneau (T.)	1,000,000

Table. 26. Weights of Metals

Name of Metal	Pounds/Cu. Ft.
Aluminum	166
Brass	504
Copper	550
Iron	450
Lead	712
Silver	655
Steel	490
Tin	458
Zinc	437

Table 27. Ultimate Bending Strength

Material	PSI
Cast iron	10,000
Wrought iron	45,000
Steel	65,000
Stone	1,200
Concrete	700
Ash	8,000
Hemlock	3,500
Oak, white	6,000
Pine, white	4,000
Pine, yellow	7,000
Spruce	3,000
Chestnut	4,500

Table 28. Recommended Lighting Levels

Area	Foot-Candles
Permieter of building	5
Office areas	70
Corridors, elevators and stairways	20
Toilets and washrooms	30
Entrance lobbies	10
Dining areas	20
Mechanical rooms	20

STEAM

Table 29. Steam Pressure Temperature Relationship

GAGE PSI	SAT TEMP F
0	212
5	228
10	240

Table 29. (Con't.)

GAGE PSI	SAT TEMP F
20	259
30	274
40	287
50	298
60	308
70	316
80	324
90	331
100	338
200	288
300	422
400	448

Table 30. Steam Trap Selection

Characteristic	Inverted Bucket	Thermostatic
Method of Operation	Intermittent	Intermittent
Steam Loss	None	None
Resistance to Wear	Excellent	Good
Corrosion Resistance	Excellent	Excellent
Resistance to Hydraulic Shock	Excellent	Poor
Vents Air and CO_2 at Steam Temperature	Yes	No
Ability to Vent Air at Very Low Pressure	Poor	Excellent
Ability to Handle Start-up Air Loads	Fair	Excellent
Operation Against Back Pressure	Excellent	Excellent
Resistance to Damage from Freezing	Poor	Excellent

Table 30. (Con't)

Characteristic	Inverted Bucket	Thermostatic
Ability to Purge System	Excellent	Good
Ability to Operate on Very Light Loads	Good	Excellent
Responsiveness to Slugs of Condensate	Immediate	Delayed
Ablity to Handle Dirt	Excellent	Fair
Comparative Physical Size	Large	Small

TEMPERATURE

Table 31. Color Scale of Temperature

Color	Temperature
Incipient red heat	900-1100
Dark red heat	1100-1500
Bright red heat	1500-1800
Yellowish red heat	1800-2200
Incipient white heat	2200-2600
White heat	2600-2900

Table 32. Centigrade/Fahrenheit Scale

°C	°F
−50	−58
−40	−40
−30	−22
−20	−4
−10	14
0	32
10	50
20	68
30	86

Table 32. Centigrade/Fahrenheit Scale

°C	°F
40	104
50	122
60	140
70	158
80	176
90	194
100	212
110	230
120	248
130	266
140	284
150	302
160	320

Table 33. Melting Point of Common Substances

Metal	Symbol	Degrees F
Aluminum	Al	1218
Copper	Cu	1981
Iron	Fe	2795
Lead	Pb	621
Mercury	Hg	−38
Molybdenum	Mo	4750
Silicon	Si	2590
Silver	Ag	1761
Tin	Sn	449
Tungsten	W	6100
Zinc	Zn	787

Table 34. Temperature of Waste Heat Gases

Source of Gas	Temperature, Deg. F
Ammonia oxidation process	1,350 - 1,475
Annealing furnace	1,100 - 2,000

Table 34. (Cont'd)

Source of Gas	Temperature, Deg. F
Black liquor recovery furnace	1,800 - 2,000
Cement kiln (dry process)	1,150 - 1,350
Cement kiln (wet process)	800 - 1,100
Coke oven	
beehive	1,950 - 2,300
by-product	up to 750
Copper refining furnace	2,700 - 2,800
Copper reverberatory furnace	2,000 - 2,500
Diesel engine exhaust	550 - 1,200
Forge and billet heating furnace	1,700 - 2,200
Garbage incinerator	1,550 - 2,000
Gas benches	1,050 - 1,150
Glass tanks	800 - 1,000
Heating furnace	1,700 - 1,900
Malleable iron air furnace	2,600
Nickel refining furnace	2,500 - 3,000
Petroleum refinery still	1,000 - 1,100
Steel furnace, open hearth	
oil, tar, or natural gas	800 - 1,100
producer gas-fired	1,200 - 1,300
Sulfer, ore processing	1,600 - 1,900
Zinc fuming furnace	1,800 - 2,000

WATER

Table 35. Water Equivalents

U.S. Gallons	x 8.33	= Pounds
U.S. Gallons	x 0.13368	= Cu. Ft.
U.S. Gallons	x 231.	= Cu. Ins.
U.S. Gallons	x 3.78	= Litres
Cu. Ins. of Water (39.2°)	x 0.03613	= Pounds
Cu. Ins. of Water (39.2°)	x 0.004329	= U.S. Gals.

Table 35. (Cont'd)

Cu. Ins. of Water (39.2°)	x 0.574384	= Ounces
Cu. Ft. of Water (39.2°)	x 62.427	= Pounds
Cu. Ft. of Water (39.2°)	x 7.48	= U.S. Gals.
Cu. Ft. of Water (39.2°)	x 0.028	= Tons
Pounds of Water	x 27.72	= Cu. Ins.
Pounds of Water	x 0.01602	= Cu. Ft.
Pounds of Water	x 0.12	= U.S. Gals.

Table 36. Supply Line Sizes for Common Fixtures

Laundry Tubs	½ inch
Drinking Fountains	3/8 inch
Showers.	½ inch
Water-Closet Tanks	3/8 inch
Water-Closets (with flush valves. .	1 inch
Kitchen Sinks	½ inch
Commercial-Type Restaurant	
Scullery Sinks	½ inch

Table 37. Pressure of Water

One foot of water = 0.4335 psi
One psi = 2.31 foot of water

Feet Heat	Pressure PSI
10	4.33
15	6.49
20	8.66
25	10.82
30	12.99
35	15.16
40	17.32
45	19.49
50	21.65
55	23.82

Table 37. (Cont'd)

Feet Heat	Pressure PSI
60	25.99
70	30.32
80	34.65
90	38.98
100	43.31
200	86.63
300	129.95
400	173.27

Table 38. Water Requirements of Common Fixtures

Fixture	Cold, GPM	Hot, GPM
Water-closet flush valve	45	0
Water-closet flush tank	10	0
Urinals, flush valve	30	0
Urinals, flush tank	10	0
Lavatories	3	3
Shower, 4-in. head	3	3
Shower, 6-in. head and larger	6	6
Baths, tub	5	5
Kitchen sink	4	4
Pantry sink	2	2
Slop sinks	6	6

Table 39. Blowdown Flow Rates – 3″ Pipe

PSI	GPS
15	0.50
20	0.60
30	0.73
40	0.86

Table 39. (Cont'd)

PSI	GPS
50	0.96
60	1.06
70	1.13
80	1.20
90	1.30
100	1.36

Table 40. Tank Capacities Per Foot of Depth

Diameter in Feet	Gallons
1	5.84
2	23.43
3	52.75
4	93.80
5	146.80
6	211.00
7	287.00
8	376.00
9	475.00
10	587.00
12	845.00
14	1,150.00
16	1,502.00
18	1,905.00
20	2,343.00

Table 41. Efficiency Loss Due to Scale

Thickness in Inches	Percent Loss
1/64	4
1/16	11
1/8	18
3/16	27
1/4	38
3/8	48
1/2	60

Glossary

ABSOLUTE PRESSURE - atmospheric pressure added to gage pressure

ABSOLUTE TEMPERATURE - the theoretical temperature when all molecular motion of a substance stops. minus 460 degrees Fahrenheit

ACCESS FLOORING - a raised floor consisting of removable panels under which ductwork, wiring and pipe runs are installed

ACID CLEANING - a process in which dilute acid, used in tandem with a corrosion inhibitor, is applied to metal surfaces for removing foreign substances too firmly attached.

ACOUSTICAL CEILING - a ceiling composed of tiles having sound-absorbing properties

AHU - air handling unit

AIR CHANGES - the number of times in an hour that a volume of air filling a room is exchanged

ALGAE - a form of plant life which causes fouling in water system piping; especially in cooling towers

ALKALINE - a condition of liquid, opposite from acidic on the pH scale, which is represented by carbonates, bicarbonates, phosphates, silicates or hydroxides contained within it.

AMBIENT TEMPERATURE - the temperature of the air immediately surrounding a device

ANEMOMETER - an instrument used for measuring air velocity

ANTHRACITE COAL - a dense coal known for its low volatility which enables the use of smaller combustion chambers than those needed to burn bituminous coal.

ASTRAGAL - a molding or strip used to cover the joint where two doors meet

ATMOSPHERIC PRESSURE - the weight of the atmosphere measured in pounds per square inch

ATOMIZATION - the process of reducing a liquid into a fine spray

AWG - American wire gauge

AXIAL FAN - a device which discharges air parallel to the axis of its wheel

BACKING PLATE - a steel plate positioned behind a welding groove to confine the weldment and assure full penetration

BACKWASH - the backflow of water through the resin bed of a water softener during the cleaning process

BAFFLE - a structure or partition used for directing the flow of gasses or liquids

BAGASSE - the dry pulp remaining from sugar cane after the juice has been extracted; a fuel used in boiler furnaces

BALANCED DRAFT - a fixed ratio of incoming air to outgoing products of combustion

BDC - bottom dead center; when a piston is at the bottom of its stroke

BEARING WALL - a wall structure which supports floors and roofs

BHP - brake horsepower; the actual power produced by an engine

BIOCIDE - a substance that is destructive to living organisms that is used in refrigeration systems by design

BITUMINOUS COAL - a soft coal generally more volatile and requiring larger combustion chambers in which to burn than anthracite coal

BLISTER - a raised area on the surface of metal caused by over-heating

BLOWBACK - the difference in pressure between when a safety valve opens and closes

BLOWDOWN - the removal of water from a boiler in lowering its chemical concentrations

BOILER HORSEPOWER - the evaporation of 34.5 pounds of water per hour from a temperature of 212F into dry saturated steam

BOILING OUT - a process whereby an alkaline solution is boiled within a vessel to rid its interior of oil or grease

BOILING POINT - the temperature at which a liquid is converted to a vapor corresponding to its pressure

BOYLES LAW - a law of physics dealing with variations in gas volumes and pressures at constant temperatures

BREECHING - a large duct used for conveying gasses of combustion from a furnace to a stack

BRITISH THERMAL UNIT (BTU) - a unit measurement of heat.

the amount of heat needed to raise the temperature of one pound of water, one degree Fahrenheit

BUS - vertical and horizontal metal bars which distribute line-side electrical power to branch circuits

BUSHING - a removable sleeve inserted or screwed into an opening to limit its size

BUTTERFLY VALVE - a throttling valve consisting of a centrally hinged plate that can be opened partially or expose the full cross section of the pipe it feeds by maneuvering the valve through a quarter turn.

BX - electrical cable wrapped in rubber with a flexible steel outer covering

CALORIE - the quantity of heat needed to raise the temperature of one gram of water, one degree centigrade

CAPILLARY ACTION - the capacity of a liquid to be drawn into small spaces

CARRYOVER - a condition whereby water or chemical solids enter the discharge line of a steam boiler

CASING - the outer skin or enclosure forming the outside of an appliance

CAVITATION - the formation of vapor pockets in a flowing liquid

CFM - cubic feet per minute

CH RATIO - carbon-hydrogen ratio

CHASSIS - the frame or plate on which the components of a device are mounted

CHIMNEY EFFECT - the tendency of air to rise within confined vertical passages when heated

CHLORINATION - the addition of the chemical chlorine to water

C I D - cubic inch displacement

COAGULATION - the initial aggregation of finely suspended matter by the addition of floc forming chemical or biological process

COEFFICIENT OF HEAT TRANSMISSION (U) - the amount of heat measured in Btu's transmitted through materials over time. the heat transmitted in one hour per square foot per degree difference between the air inside and outside of a wall, floor or ceiling.

COEFFICIENT OF PERFORMANCE (COP) - the ratio of work performed to the energy used in performing it

COMBUSTION EFFICIENCY - the ratio of the heat released from a fuel as it burns, to its heat content

COMFORT ZONE - a range of temperature and humidity combinations within which the average adult feels comfortable

CONCENTRATIONS - the number of times that dissolved solids increase in a body of water as a one-to-one multiple of the original amount due to the evaporation process

CONDENSATION - the process of returning a vapor back to its liquid state through the extraction of latent heat

CONDUIT - a pipe, tube or tray in which electrical wire are run and protected

CONTINGENCY PLANNING - a process which anticipates and prescribes corrective action to be taken in the event of unforeseen circumstances and emergency situations

CONTINUOUS BLOWDOWN - a process whereby solids concentrations within a body of water are controlled through the constant removal and replacement of the water

CONVECTION - a process of heat transfer resulting from movement within fluids due to the relative density of its warmer and cooler parts

CORROSION - the wasting away of metals due to physical contact with oxygen, carbon-dioxide or acid

COUNTERFLASHING - a downward turned flashing which overlaps an upward turned flashing used for protecting against the entry of water into a structure

COUNTERFLOW - a method of heat exchange that brings the coldest portion of one moving fluid into contact with the warmest portion of another

CPU - central processing unit. that portion of a computer which contains the arithmetic and logic functions which process programmed instructions

CRUDE OIL - unrefined petroleum in its natural state as it comes from the ground

CURING COMPOUND - a liquid is sprayed onto new concrete to prevent premature dehydration

DAMPER - a mechanism used to create a variable resistance within a gas or air passage in order to regulate its rate of flow

DB - decibel. unit for describing noise level

DEAERATION - the removal of entrained air from a liquid

DEGREE DAY - a unit representing one degree of difference from a standard temperature in the average temperature of one day, used to determine fuel requirements

DEGREE OF SUPERHEAT - the difference between the saturation temperature of a vapor and its actual temperature at a given pressure

DEMINERALIZATION - deionization. the removal of ionizable salts from solution

DESICCANT - a drying agent such as silica gel or activated alumina that is used to absorb and hold moisture

DEW POINT TEMPERATURE - the lowest temperature that air can be without its water vapor condensing

DIRECT CURRENT - an electrical current which flows in only one direction

DPDT SWITCH - double pole double throw switch

DRY BULB TEMPERATURE - the temperature of the air as measured on a thermometer

DUCT FURNACE - a furnace located in the ducting on an air distribution system to supply warm air for heating

DVW - drain, vent, waste pipes

ECONOMIZER - a heat recovery device that utilizes waste heat for preheating fluids

EFFLORESCENCE - chemical salt residue deposited on the face of masonry caused by the infiltration of water into a structure

ELECTRIC IGNITION - ignition of a pilot or burner by an electric spark generated by a transformer

ELECTROLYSIS - a chemical reaction between two substances prompted by the flow of electricity at their point of contact

ELECTROSTATIC PRECIPITATOR - an electrically charged device used for removing dust particles from an air stream

EMF - electromotive force. voltage

ENTRAINMENT - the inclusion of water or solids in steam, usually due to the violent action of the boiling process

EVAPORATION - the transformation of a liquid into its vapor state through the application of latent heat

EXCESS AIR - the air supplied for the combustion process in excess of that theoretically needed for complete oxidation

FEEDWATER TREATMENT - the conditioning of water with chemicals to establish wanted characteristics

FIRE SEPARATION WALL - a wall dividing two sections of a building used to prevent the spread of fire

FLAME ROD - a metal or ceramic rod extending into a flame which functions as an electrode in a flame detection circuit

FLAME SAFEGUARD SYSTEM - the equipment and circuitry used to provide safe control of burner operation

FLAME SIMULATOR - a device used as a substitute for the presence of flame to test a flame-detection circuit

FLAME SPREAD RATING - the measure of how fast fire will spread across the surface of a material once it is ignited.

FLASHBACK - the backward movement of a flame through a burner nozzle

FLASH POINT - the minimum temperature necessary for a volatile vapor to momentarily ignite

FLOCCULANT - a chemical used to bridge together previously coagulated particles

FLUORIDATION - the addition of the chemical fluoride to water

FLOW RATE - the amount of fluid passing a given point during a specified period of time

FLUE - a pipe or conduit used for conveying combustion exhaust fumes to the atmosphere

FOAMING - the continuous formation of bubbles having a high surface tension which are hard to disengage from a surface

FORCED DRAFT - the process of moving air mechanically by pushing or drawing it through a combustion chamber with fans or blowers

FOULING - the accumulation of refuse in gas passages or on heat-absorbing surfaces which results in undesirable restrictions to flow

GAGE PRESSURE - absolute pressure minus atmospheric pressure

GPG - grains per gallon. one grain per gallon equals 17.1 ppm

GPM - gallons per minute

GRAVITY FEED - the transfer of a liquid from a source to an outlet using only the force of gravity to induce flow

GROOVING - a form of corrosion wherein a groove is formed along the length of tubes or shells

GROUND - an electrical connection made between any structure or object and the earth

HALOGENS - chlorine, iodine, bromine or fluorine

HALIDE TORCH - a device which uses an open flame for detecting refrigerant leaks

HARDNESS - a term used to describe the calcium and magnesium content of water

HEADER - a manifold to which many branch lines are connected

HEAT EXCHANGER - a device used to transfer heat from one medium to another

HEATING SURFACE - that portion of a heat exchange device which is exposed to the heat source and transfers heat to the heated medium

HERMETIC COMPRESSOR - a unit wherein a compressor and its driving motor are contained in a single, sealed housing

HERTZ - (Hz) one cycle per second

HIGH FIRE - the firing rate at which a burner consumes the most fuel thus producing the most heat

HIGH LIMIT - the maximum value at which a controller is set that if exceeded causes the shut down of a system

HIGH TEMPERATURE BOILER - a boiler which produces hot water at pressures exceeding 160 p.s.i. or at temperatures exceeding 250 degrees Fahrenheit

HORSEPOWER - (hp) a unit of power equal to 550 foot pounds per second, 33,000 foot pounds per minute or 746 watts.

HRT - horizontal return tubular boiler

HUMIDISTAT - a control device which responds to changes in the humidity of air

HYDROSTATIC TEST - a procedure in which water is used to determine the integrity of pressure vessels

IC - internal combustion

ID - inside diameter

IGNITION TEMPERATURE - the minimum temperature at which the burning process can begin for a given fuel source

INDUCED DRAFT FAN - a fan or blower located in the breeching of gas passages that produces a negative pressure in the combustion chamber causing air to be drawn through it

INERTS - non-combustible particulates found in fuel

INTERLOCK - a senser or switch which monitors the status of a required condition which causes a programmed action to occur when the condition becomes inappropriate

I/O DEVICES - a device used to convey information to or from a computer, e.g., a keyboard or printer. input/output

ION EXCHANGE - a process for removing impurities from water on the atomic level through the selective repositioning of electrons

IR DROP - voltage drop across a resistance in an electrical circuit

JUMPER - a short length of wire used to by-pass all or part of an electrical circuit

KELVIN SCALE - a temperature scale implemented in centigrade that begins at absolute zero (−273C)

KNOCKOUT - portal designed into the side on an electrical box or metal cabinet that can be easily removed to accommodate wires or piping

KVA - kilovolt amperes

LATENT HEAT OF CONDENSATION - the heat extracted from a vapor in changing it to a liquid with no change in temperature

LATENT HEAT OF EVAPORATION - the heat added to a liquid in changing it to a vapor with no change in temperature

LATENT HEAT OF FUSION - the heat added to a solid in changing it to a liquid with no change in temperature

LIGHTNING ARRESTER - a device located in an electrical circuit to protect it from the effects of lightning.

LITHIUM BROMIDE - a chemical having a high affinity for water used as a catalyst in absorption refrigeration systems

LOCKED ROTOR - a test in which a motor's rotor is locked in place and rated voltage is applied.

LOCKED ROTOR AMPS - the amperage which is apparent in a live circuit of a motor-driven device when the rotor is not moving

LOW LIMIT - the minimum value at which a controller is set that if dropped below will result in a shut down of the system

LOW PRESSURE BOILER - a steam boiler whose maximum allowable working pressure does not exceed 15 p.s.i.

LOW WATER CUTOFF - a mechanism used for shutting off the supply of fuel to a furnace when a boiler's water level falls to a dangerously low level

LPG - liquified petroleum gas

MAKEUP WATER - water added to a system to replace that which was lost during operation due to leaks, consumption, blow down and evaporation

MANOMETER - a U-shaped tube used for measuring pressure differences in air passages

MANUAL RESET - the operation required after a system undergoes a safety shutdown before it can be put back into service

MAWP - maximum allowable working pressure

MEGA - one million times

MEGOHMETER - an instrument used for evaluating the resistance values of electrical wire coverings

MHO - a unit measurement of electrical conductance

MICRON - one millionth of a meter. 1/25,400 in.

MIXING VALVE - a three-way valve having two inlets and one outlet designed specifically for mixing fluids

MODULATING FIRE - varying the firing rate with the load thereby decreasing the on-off cycling of burners

N.E.C. - National Electrical Code

NATURAL CIRCULATION - the circulation of fluids resulting from differences in their density

NC - normally closed. a relay contact which is closed when the relay coil is not energized

NITROGEN BLANKET - a technique used whereby the air space above a body of water in a vessel is filled with nitrogen to keep oxygen from coming into contact with its metal surfaces

NO - normally open. a relay contact which is open when the relay coil is not energized

NOMINAL DIMENSION - an approximate dimension. a conventional size

NPT - national pipe thread

OD - outside diameter

OHM - a unit measurement of electrical resistance

ONE PIPE SYSTEM - a system in which one pipe serves as both the supply and return main

OPEN CIRCUIT - an electrical circuit in which the current path has been interrupted or broken

ORSAT - a device used to analyze gasses by absorption into chemical solutions

OVERLOAD PROTECTOR - a safety device designed to stop motors when overload conditions exist.

OXYGEN SCAVENGER - a chemical treatment such as sulfite or hydrazine used for releasing dissolved oxygen from water

PACKAGE BOILER - one that is shipped from the assembly plant completely equipped with all the apparatus needed for its operation

PANEL HEATING - a method whereby interior spaces are heated by pipe coils located within walls, floors and ceilings

PE - pneumatic/electric relay

PERFECT COMBUSTION - the complete oxidation of a fuel using no excess air in the combustion process

PERIPHERAL DEVICE - a hardware item forming part of a computer system that is not directly connected to but supports the processor

PF - power factor

pH - a value that indicates the intensity of the alkalinity or acidity of a solution

PILOT - a small burner used as an igniter to light off a main burner

PLENUM CHAMBER - a compartment to which ducts are connected enabling the distribution of air to more than one area

POLY-PHASE MOTOR - an electric motor driven by currents out of phase from circuit to circuit

POSITIVE DISPLACEMENT - an action wherein the total amount of a fluid being transferred by a mechanical device is accomplished without leakage or back siphonage

POTENTIOMETER - a variable resistor in an electrical circuit

POWER FACTOR - an efficiency value assigned to electrical circuits based on a comparison of its true and apparent power characteristics

PPM - parts per million

PRECIPITATION - the removal of constituents from water by chemical means. condensation of water vapor from clouds

PRESSURE REGULATOR - a mechanism used to maintain a constant pressure within a feeder line regardless of fluctuations above the setting in the supply line

PRIMARY AIR - combustion air which is introduced into a furnace with the fuel

PRIMING - the discharge of water particles into a steam line

PRODUCTS OF COMBUSTION - any gas or solid remaining after the burning of a fuel

PRV - pressure-reducing valve

PSI - pounds per square inch

PSYCHROMETRIC CHART - a graph which depicts the relationship between the pressure, temperature and moisture content of air

PURGE - eliminating a fluiɑ from a pipe or chamber by flushing it out with another fluid

PVC - polyvinyl chloride

RADIATION LOSS - the loss of heat from an object to the surrounding air

RECTIFICATION - the conversion of alternating current to direct current

REFRACTORY - heat-resistant material used to line furnaces, ovens and incinerators

REFRIGERATION - the removal of heat from an area where it is not wanted to one that is not objectionable

REGISTER - the grill work or damper through which air is introduced

RELAY - an electromechanical device having a coil which, when energized and de-energized, opens and closes sets of electrical contacts

RELIEF VALVE - a device used to relieve excess pressure from liquid-filled pressure vessels, pipes and hot water boilers

RH - relative humidity

RINGELMANN CHART - a comparator of smoke density comprised of rectangular grids filled with black lines of various widths on a white background

RMS - root mean square

RUNNING CURRENT - the amperage draw noted when a motor is running at its rated speed

SBI - Steel Boiler Institute

SAIL SWITCH - a switch attached to a sail-shaped paddle inserted into an air stream which is activated when the air stream striking the sail reaches a pre-established velocity

SATURATED STEAM - dry steam which has reached the temperature corresponding to its pressure

SCALE - a hard coating or layer of chemical materials on internal surfaces of pressure vessels, piping and fluid passages

SEDIMENTATION - settling out of particles from suspension in water

SENSIBLE HEAT - heat which changes the temperature but not the state of a substance

SET POINT - a pre-determined value to which a device is adjusted that, when reached, causes it to perform its intended function

SHORT CIRCUIT - an unintentional connection between two points in an electrical circuit resulting in an abnormal flow of current

SLING PSYCHROMETER - a device having a wet and a dry bulb thermometer which measures relative humidity when moved rapidly through the air

SNG - snythetic natural gas

SPALLING - deterioration of materials evidenced by flaking of their surfaces

SPECIFIC GRAVITY - the ratio of the weight of any substance to the same volume of a standard substance at the same temperature

SPECIFIC HEAT - a measure of the heat required in Btu's to raise the temperature of one pound of substance, one degree Fahrenheit. the specific heat of water is 1.0

SSU - Seconds Saybolt Universal

STATIC HEAD - the pressure exerted by the weight of a fluid in a vertical column

STATIC PRESSURE - the force exerted per unit area by a gas or liquid measured at right angles to the direction of flow

STRATIFICATION OF AIR - a condition of the air when little or no movement is evident

SUMP - a container, compartment or reservoir used as a drain or receptacle for fluids

TDC - top dead center. when a piston is at the top of its stroke

TDS - total dissolved solids

TENSILE STRENGTH - the capacity of a material to withstand being stretched

TERRAZZO - a material used for poured floors consisting of concrete mixed with marble chips

TERTIARY AIR - air supplied to a combustion chamber to supplement primary and secondary air

THERMISTOR - a solid state device whose electrical resistance varies with temperature

THERMOCOUPLE - a mechanism comprised of two electrical conductors made of different metals which are joined at a point which, when heated, produces an electrical voltage having the value that is directly proportional to the temperature of the heat being applied

THERMOPILE - a battery of thermocouples connected in series

THERMOSTATIC EXPANSION VALVE - a control device operated by the pressure and temperature of an evaporator which meters the flow of refrigerant to its coil

TON OF REFRIGERATION - the heat required to melt a one-ton block of ice in 24 hours. 288,000 Btu's or 12,000 Btu's per hour

TWO POSITION VALVE - a valve which is either fully open or fully closed, having no positions in between

UNDERCARPET WIRING SYSTEM - flat insulated wiring designed for running circuits under carpeting where access to wire chases under the floor are not available

UNINTERRUPTIBLE POWER SUPPLY - a separate source of electricity used to maintain continuity of electrical power to a device or system when the normal supply is interrupted

VACUUM - any pressure less than that of the surrounding atmosphere

VAPOR RETARDER - a barrier constructed of materials which retard the capillary action of water into building structures

VAR - volt-ampere reactive

VAV - variable air volume

VELOCIMETER - an instrument used to measure the speed of moving air

VENTURI - a short tube designed with a constricted throat that increases the velocity of fluids passed through it.

VISCOSITY - a measure of a fluid's resistance to flow

WET BULB TEMPERATURE - the lowest temperature that can be attained by an object that is wet and exposed to moving air

Index